# Splendid Was the Trail

## by Kenneth D. Swan

With Photographs by Kenneth D. Swan
Vignettes by Joseph S. Swan

Introduction by Arnold Bolle

Richard B. Roeder
Series Editor

*Sweetgrass Books*

Published by
Montana Magazine
American & World Geographic Publishing

Library of Congress Cataloging in Publication Data
Swan, Kenneth D. (Kenneth Dupee), 1887-1970.
    Splendid was the trail / by Kenneth D. Swan ; with photographs by Kenneth D. Swan ; vignettes by Joseph S. Swan ; introduction by Arnold Bolle.
        p.      cm — (Sweetgrass books)
    Originally published: Missoula, Mont. : Mountain Press, 1968.
    Includes index.
    ISBN 1-56037-035-1
    1. Swan, Kenneth K. (Kenneth Dupee), 1887-1970. 2. Forest rangers—Montana—Biography. 3. United States. Forest Service. Northern Region—Officials and employees—Biography. I. Series.
SD129.S97A3  1993
634.9'092—dc20
[B]                                                                    93-21346

To my wife, Ruth Barrows Swan

# Acknowledgments

*From the First Edition*

Incidents in connection with the trip in the Mission Mountains with Jack Clack were used as the basis for a story in diary form which appeared in *The Living Wilderness* magazine in the winter issue of 1965-66. I wish to thank the *Wilderness Society* for permission to use this copyrighted material. The article has been completely re-written, but preserves the essentials of the original account.

I wish to thank the Montana Forestry Club of the School of Forestry of the University of Montana for permission to use an article on photographic points which I prepared for the 1961 issue of the Forestry Kaimin.

My thanks go to those old friends who have helped me verify material used—Ed Mackay, Flemming Stewart, Stacy Eckert, Earl Cooley, Ralph Hand, John Taylor, Burr Hurwitz, Lloyd Noel, and others.

Lastly I wish to acknowledge the valuable help given me by Dr. Edwin S. Leonard, Jr., Dean Emeritus of Principia College, in editing the final manuscript. His counsel has made the task much easier.

All of the photographs, unless otherwise credited, were taken by the author during the years he was working as official photographer for the Northern Region of the United States Forest Service.

# Contents

# Introduction

Kenneth D. Swan's career with the United States Forest Service from 1911 until his retirement in 1947 spanned the period popularly known as the "Stetson Hat" era of the Forest Service. This was the time when the main activity of the personnel was protection of the forests from fire, disease, insects, and human incursion of various kinds. The forest ranger was pictured in uniform on horseback wearing an appropriate Stetson hat: a highly respected man whose presence assured us that we could rest easy; our forests and all the living things within its borders were safe.

K.D. (as he was popularly called) was best known for his pictures, the photographs he took throughout his career. They illustrated the many talks he gave around the region as part of his duties in the Information and Education branch of the northern Rockies regional office. His task was education: of school children, the general public, and many groups and organizations. K.D. was perfectly suited to the job and became a featured speaker. One didn't have to live long in Montana before encountering a K.D. program. His life became devoted to the issue of public education. His pictures were, and still are, works of art that thrill his audiences.

The devotion K.D. held for the forests and for the mountains was communicated beautifully. He pioneered with film, with cameras, and with effects. His talks fitted the pictures and, with this book, he shows that he was also a skilled writer. He was a gentle man who loved the West. He became a true Westerner in thought, though he was a fine gentleman as well as a gentle man.

The book tells his story effectively and tastefully. Helen—his daughter and my wife—and I believe that much of his gentle nature is conveyed by his language and by the pictures he paints in this account. We can be thankful for this reprint by Montana Magazine. K.D. had a second volume underway when he died. There is far more to tell and it's too bad he didn't finish it. The story is timeless and actually increases in interest and value as this period of forestry fades into the background and tends to be forgotten under the violence of the "Hard Hat" era that succeeded it.

Kenneth Dupee Swan arrived in Missoula in the summer of 1911 with a brand new master's degree in forestry from Harvard University. He was one of the new professional foresters to join the U.S. Forest Service after the great forest fires of 1910. Those fires had alarmed the nation and brought about increased funding to prevent such disasters from recurring.

The Forest Service undertook a massive drive to attract major timber industries into the area to salvage the billions of board feet of scorched timber, but to no avail. There was still plenty of cheap timber available in better growing sites closer to markets, so the new foresters were put to work at other needed tasks. One of these was surveying and map-making. There was a need to establish boundaries for the newly created National Forests as well as to sketch out the various features of the landscape, such as rivers, mountain peaks and ridges, and to delineate and describe vegetative types, especially good stands of timber. Amid the many tales of desolation and heroism of the 1910 burn, this might have been considered rather drab work, but K.D. took to it and delighted in the mountains and wild country.

In a brief foreword, K.D. tells of his early-day rambles in the vicinity of Boston, and of finding wild country there. He doesn't mention the fact that he spent summers in the mountains of New Hampshire as well, including a summer in charge of "The Hut" on top of Mount Washington. He had developed a strong link to mountains and wilderness before he came to Montana. Here it was love at first sight and it lasted all of his life.

He also loved the people he met or worked with along the trail, starting with Lou Morgan, ranger at the Neihart station. There regional (then district) forester Fred Silcox assigned K.D. He describes Morgan as "a kindly man who was well-liked by all." K.D.'s task was surveying homesteads claimed along stream bottoms within National Forest boundaries newly permitted under the "Forest Homestead Act" of 1911.

K.D. assembled his camping gear, bought a saddle horse, a saddle for $20, and his first Stetson hat for $5. He still owned and used the saddle and the Stetson when he retired almost 40 years later, which speaks clearly of K.D.'s meticulous care for his equipment as well as his thrifty New England nature. His Stetson became famous by its appearance in photos later on, as a measure of size or place.

With Lou Morgan and a helper, gear and luggage was packed into a wagon and the party set off into the Musselshell Valley, stopping off at stations, meeting homesteaders, surveying and mapping as they went. K.D. loved the "broad scope" of the countryside that gave him his first real encounter with the "great open spaces." All were sparsely populated with especially fine people. John Bonham, ranger at the Musselshell Ranger Station, was a "wonderful man," a true Westerner always decked out for the saddle. His wife was an "excellent cook" possessed of other fine qualities. She looked after the office as well as the kitchen, "typical of the early-day ranger's wife," a true helpmeet.

K.D. met his first sourdough hotcakes at the Haymaker Ranger Station. From that day forward, he was never without "starter" for the rest of his life. In this book, he describes how to make them. Helen and I were amazed to read this story, because when he provided us with some of his starter he told us it had come over on the *Mayflower*.

He found the homesteaders fine, friendly people, and was happy with his first summer's work. Later, during the fall, K.D. was assigned to one of the first tree-planting attempts in Dry Coulee in the Snowies. There his horse, Tony, and his pack horse, Sailor, became "indispensable" in packing planting stock to heeling-in beds for the planting crews. His late November return over the mountains to Neihart gave K.D. his first rugged blizzard experience, which he describes in detail. In the tough winds and bitter cold he saw and lived the West at its wildest and he relished every minute of it.

For his first winter K.D. was brought into the Missoula regional headquarters to help work piles of field notes into yield-tables for timber. He was an aide to John Pristine, "one of the ablest in the Forest Service." K.D. remarks that he much preferred field work and was pleased to learn that his assignment for the next summer was to Camp Crook on Montana's eastern border.

The Sioux National Forest consisted of scattered hills and buttes with a fringe of timber on top. Chalk Buttes, Ekalaka Hills, and Lone Pine Hills in Montana, and Short Pine Hills, Slim Buttes and Cave Hills in the Dakotas supplied sorely needed timber for homesteaders settling in to break up native sod and hopefully make their fortunes in growing wheat. They needed timber for cabins, corrals, barns, fences, and some fuel (though that sometimes came from lignite coal), which helped keep settlers warm during the long windy winters. K.D. found the country, as well as the people, charming. He scaled timber, and made topographic maps with the aid of aneroid barometers. There wasn't much timber, and what there was pretty scrubby stuff. Here the demand for timber provided the strongest market for forest products in the whole region.

The social life was active too. The "honyockers" were mostly young and active people who had plenty of energy for dancing and visiting each other. It's hard now to imagine all that life there.

After another winter at a desk in the regional headquarters, K.D.'s next assignment was to the heavily timbered white pine forests of Idaho's Clearwater Forest. Timber cruising and topographic mapping occupied the summer, followed by another winter in Missoula compiling data and drafting field maps.

In the mid-1920s, the Forest Service established a division of Information and Education to take its message to the public. K.D. was assigned to the new division, which was headed by Ted Shoemaker. K.D.'s account of the first "showboat" trip to eastern Montana with Ted Shoemaker gives a highly interesting description of living conditions In that area then. Their conveyance was a Model T pickup and the roads, never paved, were often mere ruts in the prairie. They brought their own generator because some of the 28 towns they visited had no electricity. Their show included a two reel movie and slides, and drew packed houses wherever they went. People in those days thirsted for entertainment and happily soaked up the message and the pictures, most of them K.D.'s own.

Whether part of the education mission or not, K.D. and Shoemaker started the Mountaineers Club in Missoula. This group took a trip into the backcountry each weekend and hosted potlucks on Friday evenings. Advised of it by fellow students, this was the first group I joined when I came to Missoula in 1937. Mountains were new to me and I set out to climb every one I could see. The potlucks were manna from heaven for a starving student. K.D.'s slide talks were a regular feature of the evening gatherings. To me the Mountaineers still mean a lot for it was there I met my future wife, K.D.'s daughter Helen.

The more pictures K.D. took, the more his fellow workers wanted him to take. Before long, major photographic assignments came his way. The first of these was fire, which was the main concern of the Forest Service in the northern Rockies at that time. K.D recounts early days on backcountry fires. There were few maps and few trails. Fire crews and pack strings took days to get to the fire and sometimes got lost on the way. He tells of the development of highly efficient pack strings, trained to enter trucks, which sped them to the nearest roadhead. He goes on to tell of the beginnings of smoke jumping and equipment development centers. During K.D.'s time, getting to the fire and having it under control by 10 A.M. of the next day was the rule. There was as yet no thought of letting fires burn.

The account is livened with vivid personal accounts of lost crews and fire crews kept on after the fire to build trails. All of it is brought to life with amazing pictures carefully taken and painstakingly developed into works of art. K.D.'s reputation grew. While his earlier pictures were labeled just "Forest Service Photo," the name K.D. Swan started being added, an almost unheard of distinction for a government employee.

He was a careful photographer. Everything had to be just right. He experimented with various cameras but his favorite from the start was a "cumbersome view camera." "For me," he writes, "there was immense satisfaction in pulling a black cloth over my head and seeing the image on the ground glass viewing screen." In this way he could "judge the merits of the composition. I often moved the camera several times in composing a picture. Patience paid big dividends...." The equipment weighed over thirty pounds. Jack Clack, another forester and a close friend, designed two special pack boxes to contain and protect the equipment. The pack boxes still exist in perfect condition, of course. Smoke Elser, the outfitter, now has them on display at his place in the Rattlesnake Valley.

Old-timers delight in recounting photo trips with K.D. He was very careful setting up. He spent hours just looking over the scene and deciding what to do, where and when. Monk Dejarnette told of such a trip with

K.D. on which, after two days of studying the scene, K.D. just decided it was no good. They packed up and came back without a single photograph. K.D. mentions the picture of Fort Union, where he decided to take just one shot. He didn't believe in wasting film. Recently a photographer was sent out for a picture of me. He talked and snapped oodles of pictures, hundreds of them, and he needed only one. I told him about K.D.'s approach. His response was like this: "In those days they carefully took a picture. Now we just snap a lot of prints hoping among them there will be a picture."

The life of the photographer is extremely interesting for the reader. His photo assignments took K.D. into many interesting areas. His pictures of the Civilian Conservation Corps, of more fires, and of many Forest Service activities form a vivid picture history. His discovery of "photopoints" by retaking a picture of his Stetson hanging on a pine tree after a twenty-year interval, opened a new field of opportunity; there is now a grid of such photo points and a growing file of pictures to depict changes in forest growth over time.

K.D.'s favorite picture-taking was in the wild backcountry. Well before the Wilderness Act, he was taking pack trips into these areas, getting pictures that were widely used to inform people of the attractions of these places. The railroad companies had been urging the agency to publicize trips into the wild country in order to attract more passengers on their lines. Whatever the reason, K.D. treasured those opportunities to camp out by the mountain tarns and vivid cliffs. He took many of his best photos here and his love for the wilderness comes out strongly. You can almost feel his devotion in the pictures he took.

The book ends with K.D.'s retirement in 1947, just before the start of the postwar building boom. He grew increasingly unhappy over the spread of roads and clearcuts throughout the region. He took no action and seldom spoke of it but you could read the unhappiness in his expression. The Stetson Era was over and he was sad. He continued taking pictures. His life followed a pattern. Here in Montana for the summer, a trip to New England for the fall colors each autumn, and winter in Wickenberg, Arizona for spectacular photos of the desert. He expanded his slide shows and continued to be in high demand as a speaker wherever he went.

Through the efforts of John Pierce of Missoula, K.D.'s pictures have been carefully cataloged, his slide shows taped and automated, and donated to the files of the University of Montana library archives. They are available on loan to the public. These are his personal pictures. The official Forest Service photos and negatives are on file in government archives in Washington, D.C. Judd Moore, of the Forest Service's regional office, Missoula, sent for a list at my request and the list was staggering. Catalogued only by a mysterious code, they are impossible to identify by viewing them individually. I tried that back In Washington one day, and found pictures of pine cones, of various tools and equipment, and only an occasional picture of scenery or interesting people activity. Sorting this collection would be staggering, but what a collection it could be!

K.D.'s story of his "trail" speaks only of his official work, nothing of his personal life. He returned to Boston after his first summer in Montana and married a woman he had met in North Conway, New Hampshire, while he worked in the White Mountains. His son, Joe, was born in Lewiston, Idaho, and his daughter, in Missoula, two years later. Helen tells of many happy picnics and camping out in the countryside near Missoula. She also tells of her mother's lingering illness and death at an early age. K.D. does mention his second wife as his companion on several trips into the mountains in connection with his work. His private life is never mentioned otherwise.

K.D. died in March of 1970 at the age of 81. He had thoroughly savored his "splendid trail," and left a valuable legacy in word and image.

*Arnold W. Bolle*
*Missoula, Montana*

# The Way Beckons

My first love was a range of hills—the Blue Hills of Milton and Quincy—high points on the south rim of the Boston Basin. As a youngster I gazed from my home to where in the south they gave character to a distant sky-line, rising to no great heights, but having to my boyish fancy the semblance of a mountainous terrain. They were granite hills, worn to gentle contours and polished by the great ice sheet which once covered the region. From them had come the stone for the monument on Bunker Hill, the ponderous granite blocks being transported by horses and oxen from the quarry to a wharf on the Neponset River, where they were loaded onto barges for the journey across Boston Harbor to Charlestown, site of the memorial obelisk.

Great Blue, at the western end of this range of hills, rises to an elevation of 635 feet above sea level, and is considered the highest point near the Atlantic Coast south of Maine. A famous meteorological observatory was built on the summit in 1884, and named for its sponsor, A. Lawrence Rotch. Stretching for several miles to the east from here are lesser summits; a galaxy of rough, broken hills which terminates abruptly at Babel Rock in West Quincy. Most of the hilly area, roughly six miles long by three miles wide, has been for many years under the jurisdiction of the Metropolitan Park Commission of the State of Massachusetts.

This was my wilderness, easily reached from home by bicycle, trolley, or even on foot, and here I spent countless carefree hours in all seasons—following trails and old wood roads, climbing to rocky summits, discovering abandoned quarries, and exploring woodsy hollows seldom disturbed by human footsteps. I knew well the bogs where the peepers would raise the first frog chorus in spring; where to find hepaticas and lady's slipper orchids in season; places where the swamp maples would first turn to crimson in autumn. I learned to identify the shrubs and trees which I met on my rambles; here in the hills I became an ardent bird-watcher.

Here, too, was born a desire to become a forester. At that time I knew little about the demands of the profession, but I did know that I wanted work which would keep me in the open most of the time: life behind a desk was not for me, so I reasoned. My decision was made before I was graduated from high school—a decision I have never regretted. Five years later, with a degree in forestry from Harvard University in one hand, and an appointment as Forest Assistant in the United States Forest Service in the other, I took a train for Missoula, Montana, my assigned headquarters. That was the beginning of a trail which in thirty-seven years gave me an intimate acquaintance with many western national forests. Those were happy years—the years when I was active in the United States Forest Service—years to look back upon with deep satisfaction.

# Little Belt Days

On July 1, 1911, P.J. O'Brien, legal advisor for District 1 of the Forest Service, administered the oath of office to twenty new forest assistants who had managed to crowd into his tiny room in the old Hammond Building which stood on the site of the present Hammond-Arcade complex. After this ceremony the group was briefed by District Forester Silcox, and each man assigned to the forest where he would work. I was instructed to report to the supervisor of the Jefferson National Forest (now part of the Lewis and Clark Forest) in Great Falls, Montana, and told that he would probably send me to Neihart to help the district ranger, Lou Morgan, during the summer.

At this time the once prosperous silver camp of Neihart was well on its way to becoming a ghost town. Many stores and saloons along the main street had been boarded up, the plank sidewalks were badly in need of repair, roofs of deserted cabins were falling in. A few miners lingered to busy themselves with the sporadic operation of one of the mines which produced a dribble of silver ore to keep alive the hope that sometime Neihart would "come back." This handful of miners found board and lodging at Mother Mix's boarding house, or batched in shacks which had seen better days.

A train made its way on weekdays from the main line of the Great Northern at Armington up the tortuous valley of Belt Creek to a terminus at the forlorn little settlement. This was a slow stub, running mostly late, which served communities and whistle stops along the way. The town of Monarch, fifteen miles or so below Neihart, was the most important of these.

Situated as it was in the heart of the Little Belt Mountains, whose rugged, timbered terrain had been included in the newly created Jefferson National Forest, Neihart was naturally selected as the headquarters of a forest ranger district. A snug log building was erected a few miles down the valley, and designated the Belt Creek Ranger Station. It was here I first met Lou Morgan.

Lou was a native of the Sun River country, where members of his family had been engaged in a mercantile business. He was well-liked by all who knew him, and his faculty of getting along with people, coupled with an ability to pick up the practical knowledge necessary to make a success of the forest ranger's job, gave him a high rating on the staff of the Jefferson Forest. He was a kindly man, and took me, a green Easterner, under his wing and did everything possible to help me become adjusted to my new position as Forest Assistant.

In those days field-going men were required to provide their own horses and riding equipment, and to pay for the upkeep of their animals. I had been in the saddle only a few times in my life, and to make a wise choice of a horse posed a problem. Here Lou took over and gave me a world of sound advice. He procured for me a chunky little sorrel named Tony, for which I paid $50—a fair price in those days. Gentle and easy-gaited, Tony was just the animal for a greenhorn. Shortly afterwards Lou made a trip to Great Falls on official business and brought me back a saddle—a piece of equipment which has served me well at odd times ever since. For this I paid $35!

Congress had passed on June 11, 1906 an act whereby areas of agricultural land within the national forests were made available for homestead entry. Under this act a huge program of land classification was carried out and several million acres of land withdrawn from National Forest reserves.

Forest homesteads of this sort were generally referred to as June 11th claims, in reference to the date of the enabling act. Inasmuch as these tracts of agricultural land within the forests were generally irregular in shape, they had to be defined by metes and bounds surveys. Care was taken to exclude any sizable bodies of timber. Each homesteader was entitled to not more than 160 acres.

At the time I was assigned to the Jefferson Forest, a considerable number of these June 11th claims along the south side of the Little Belt Mountains were awaiting preliminary examination. Each tract would be surveyed with a compass and chain. These preliminary surveys were accurate enough for practical purposes, but later, after "proving up," another more refined survey was made by a crew of trained surveyors. Then, and only then, the homesteader received a deed to his property.

Lou Morgan was given charge of this preliminary work, and I was to be his assistant. James Yule, a native of the Neihart country, was a third member of our little party. Jim had a flair for mathematics, and was of immense help in juggling latitudes and departures and figuring areas.

Several days were spent in assembling equipment and supplies for our expedition. A tent, bedding rolls, a sheet iron stove, cooking utensils, supplies (including the precious kettle of sour dough starter), maps, notes, and instruments needed for the work all had to be stowed in a wagon and protected with a canvas cover. A team of well-matched horses pulled the outfit with Lou in the driver's seat. Jim and I were on horseback with slickers tied behind the saddle and cantinas bulging with knickknacks dangling from the horn. I am sure we felt rather important as our little caravan moved up the main street of Neihart that July morning, bound for King's Hill and the country to the south.

Our destination at the end of the first day's ride was a forest service cabin at a location known in those days as Wolsey. Here we met Bill Kierstead, a summer employee of the Forest Service, with quite a reputation in those parts for his energy and perseverance. Small-framed, wiry, and slightly under normal height, he was a tireless hiker; it was said he often went fifty miles on foot in a day if horses were not available. He was a kindly man, and seeing I was tired from the unaccustomed ride of twenty-five miles, he helped me care for my horse. Lou, Jim, and I found sleeping quarters in a loft, unprotected from mosquitoes which pestered us until the coolness of night brought relief.

Here seems an appropriate place to take a look at the bed of a forest officer in those days. It was often referred to as a sheepherder's bedroll. Invariably there would be one or more quilts known as sugans, the filling of which consisted of shoddy linter material. The covering was usually bright-colored calico whose colors ran badly if exposed to the weather. Sometimes there would be a shoddy blanket, or a pair of cotton flannel sheets which might in the case of the more fastidious men be washed once each season. All this was rolled in a piece of heavy canvas which served as a ground cloth, or perhaps as protection against rain or snow. Air mattress-

*Splendid Was the Trail*

es were not in general use at this time: one spread his bed on the ground or a cabin floor, with perhaps a sugan folded lengthwise to serve as padding. But sleep came quickly and was sound; one got up refreshed and ready for the day's work.

Leaving behind the lodgepole forests of the Little Belts, we traveled across open benchlands and through the Smith River and Musselshell Valleys. I was fascinated by the broadness of this land which was different from any.I had seen before. There was no feeling of being shut in such as one sometimes experiences while traveling the gulches of a mountainous terrain. Our pace was slow, and gave us ample time to ponder the landscape as it unfolded to us—to scrutinize the mountain ranges as they came into view—the Castles, the Big Belts, the Crazy Range, the Snowies. It was pleasant to jog along, relaxed in the saddle, unpressed for time. During the noon hour of the second day I took a dip in Smith River, much to the amusement of Lou and Jim. The summer sun that day was scorching hot; after coming out of the water I lay in the shade of a cottonwood and watched the horses grazing along the stream bank.

On a Sunday morning, three days out from Neihart, we drew up at the Musselshell ranger station and caught forest ranger John Bonham at home and out of the saddle. John was a wonderful man. He had all the qualities we like to ascribe to those resourceful individuals who made a success of life on the open range when settlers were few and far between. The impression one got of John as a real cow man was heightened by a drooping mustache, tall riding boots, and a lariat which he always carried on his saddle. On a national forest where grazing was an important activity, John, with his intimate knowledge of the livestock industry and the needs of the stockman, made an excellent ranger.

On John's district we began our surveying activities. On the eastern edge of the Castle Mountains was a small tract which was agricultural in character and appeared to be suitable for cultivation. By dint of hard work a settler could put in a grain crop which in a good year would yield a small income, and could acquire grazing rights on national forest land, which might well prove to be of more value than an agricultural crop. A spring or well would provide a domestic water supply, ample perhaps to irrigate a vegetable garden and a few fruit bushes. Free dead timber from the national forest could be obtained for the asking. I thought it a beautiful location for a home as we rode into the open grassy park on that summer morning. The land had not been grazed, and flowers among the tall grass were there in profusion. Notes of meadowlarks came from all directions, so it seemed.

The boundary of this tract was quite irregular because of the marginal areas of timber which we conscientiously tried to exclude in making our survey. Lou took readings with the compass, and Jim and I measured distances with the 66-foot surveyor's chain. Good-sized rocks inscribed with a cold chisel and witnessed by pits, or blazes on nearby trees, were set in the ground at the corners. Lou kept the notes with great care, from which we later figured the area of the tract, using tables of latitudes and departures from a surveyor's manual. Lou was a wizard in taking readings with a Forest Service compass, for on this and subsequent surveys his closures were remarkably accurate.

In those days the Musselshell River and tributaries in the vicinity of Delpine, where the ranger station was located, were teeming with trout. Mrs. Bonham insisted we eat at the family table during our stay with John; she was an excellent cook, and made sure we had an abundance of sourdough hotcakes in the morning and heaping platters of trout fried to golden brown crispness in the evening. Very much alive, a good conversationalist, and an efficient manager always available on the telephone at the time of a fire or other emergency, Mrs. Bonham was typical of the early-day ranger's wife.

Daisy Dean, Haymaker, Hopley—these creek names bring memories of locations where we worked during the next two months. Often the would-be settler would meet us on the ground to point out the exact piece of land he wished for a homestead. Many would tell us of their dreams for a better life than they had known in the treadmill of city employment and living, and confide that it was "now or never," as they had cut old ties and were prepared to invest a life's savings in this new venture. Truly, the prospects looked good at that time, for we were in a wet cycle and the rainfall over the past few years had been sufficient for successful dryland farming. A general spirit of optimism was abroad, created in no small measure by the promotional advertising of the newly constructed Chicago, Milwaukee, and Puget Sound Railroad. The bright picture was not dimmed by a knowledge of the drought years ahead, when diminishing rainfall and drying winds would reduce crops to a mere nothing. And yet it is only fair to say that these little ranches in the shadow of the rain-making hills probably fared better in the drought years than farms on the prairies farther east where dust bowl conditions prevailed. We liked these people and wished them well. To me, fresh from the East, these settlers, many of them Scandinavians in origin, epitomized the free spirit of the Western pioneer.

It was at the Haymaker Ranger Station that I first learned the art of making sourdough hotcakes. Our starter was kept in a five-pound lard pail, and carried tenderly in the wagon on moving days. I remember one hot August day when the over-heated batter blew the top off the pail and erupted over a box of dishes, which had to be cleaned up before they could be used for the evening meal.

It may be well to explain a little more in detail just how one cares for and manipulates this item so essential to pioneer living. To begin with, the dough, or starter, is a culture of the yeast plant. These cultures are considered very precious, and are often, one might say, handed down from father to son. The starter we were using was said to have been in Lou's family some fifty years or more. It might even have been traced back to days when the family crossed the plains in covered wagons to finally settle in the Rocky Mountains.

Usually Lou presided over the sourdough kettle, but one morning he consented to give me a lesson in cake making: "Put as much dough as you think you will need in a mixing bowl," said he. The sticky batter seemed plenty sour, and gave out a strong, yeasty smell. To this was added some salt, sugar, and a beaten egg, and canned milk enough to thin the mixture so it would spoon onto the griddle easily. Finally was added a teaspoon of baking soda moistened with a little water. "Too much and you will taste it, too little and the cakes will taste sour," said Lou. The reaction between the alkaline soda and the slightly acid batter caused the mixture to effervesce with countless tiny bubbles. Fried to a golden brown on a cast iron griddle the cakes were delicious: we ate them with syrup made from brown sugar. After breakfast, flour was added to the starter kettle to compensate for the amount of dough that had been taken out. A clean white cloth was stretched over the top of the pail and the cover pressed down firmly.

The summer of 1911 brought more than normal rainfall, and on some days we could not work outside. Luckily we were able to find shelter in Forest Service buildings during these rainy periods, and were not often obliged to pitch the tent. A new house had just been completed on Haymaker Creek, and here we set up housekeeping for a couple of weeks. One day when we were being treated to a cold easterly storm I elected to ride to Martinsdale, some twelve miles across the valley, to pick up the mail and a few grocery items which could be carried behind the saddle. Clad in Lou's long yellow slicker and a stiff-brimmed felt hat which shed water like a tin roof, I set off for a four-hour ride through a downpour which bid fair to turn into snow before evening. It was a dreary ride exposed to gusts of chilling wind with all but the closest landmarks obscured behind curtains of rain and mist. Wagon tracks and horse trails became gumbo traps where the sticky clay would build up on the horse's hooves to make travel well-nigh impossible. To progress at all it was necessary to seek out a way over unbroken sod. But, slithering and sliding, Tony and I finally reached a stretch of graveled road which finally led us to the main street of the little town.

Martinsdale seemed a busy village after the quiet of the hills where we were working. In spite of the storm and the deplorable condition of the roads, saddle horses and teams were tied to the hitchracks along the one street, and slicker-clad shoppers made their way hither and yon regardless of the rain. With the coming of the newly built railroad, the town had gained importance as a shipping point for sheep, cattle, and wool. Here was activity and an appearance of prosperity, much different from the atmosphere of decadence which greeted one in Neihart.

As I turned homeward the rain ceased, and before I reached Haymaker the sun shining through a rift in the storm-clouds to the west gave hope of a final clearing. Next morning there was a promise of fall in the nippy air: the Crazy Mountains gleaming white with new snow reminded us that summer is short indeed in this part of Montana.

On Hopley Creek, near the point where it issues from the hills in a limestone canyon, and about twelve miles due west of the town of Judith Gap, was a small two-room structure on the administrative site designated as the Muir Ranger Station. Here our party was joined by Henry Sutton, a tall fine-looking miner from the Neihart country. Henry, possibly forty years of age, was extremely popular with the ladies, but in spite of the machinations of the belles of Neihart and elsewhere, had remained a bachelor. Except for a short period in the Navy, which had taken him to far-eastern countries, he had followed mining, mostly in the Little Belts. Failing to strike it rich in his last venture, he had decided to try a shift as a forest guard, a temporary position in the United States Forest Service.

Henry rode into camp on a saddle mare, leading a tall pacer of uncertain years fitted with a sawbuck saddle on which were carried a bedroll and other personal belongings, all made fast by a diamond hitch. This bearer of burdens went by the name of Sailor, and a more willing and gentle animal never followed the trails of the Little Belt Mountains, I am sure. At a later date I became the possessor of both animal and saddle for the sum of fifty dollars. I never regretted the purchase.

On a Saturday in early September Judith Gap celebrated its third birthday. I was sent to town to get the mail, with the understanding that probably we would make up lost time on the following Sunday. There had been a copious rain, with snow on the mountains, and the day was bright and clear as I rode into town. In a few places the wheat harvest was underway, and threshing crews had been busy since daylight. The machines, cumbersome and noisy by present-day standards, were belching black smoke and building straw stacks of huge dimensions. Here wheat was king, and would surely bring an era of prosperity to this new land, so thought the busy workers toiling to get the precious grain to the elevator before wintry storms set in.

The little town was making a brave show that day with flags and bunting. People from surrounding settlements were there in numbers, determined to make the most of the brief holiday by mixing with their neighbors and exchanging the latest news with friends whom they had not seen perhaps for many months. Hitchracks were all in use, and vacant lots and other available space occupied by teams, unhitched until the hour of homegoing and calmly munching hay, or perhaps a feed of oats measured out in the wagon box. Children of all ages were everywhere, romping and racing as children will, or perhaps moving more sedately as they held the hand of a parent or grownup. Little girls, and big ones too, I seem to remember, wore their hair in pigtails. For an event of this kind they often came forth in starched dresses, with the hem line well below the knees. Bib overalls were favored by boys of all ages as well as by the men. Happy the youngster who had a nickel to spend on popcorn at the magnificent cart of rococo design pulled by a gentle little pinto horse, a great pet with the children.

On the whole it was a remarkably sober crowd. Buckaroos, some conspicuous in chaps, who would compete in the bucking contests later in the day, swaggered a little for the benefit of the children, but there was little evidence of excessive drinking, and no rough talk—in front of the women,

*Splendid Was the Trail*

at least. More than one cow hand that I saw held his ten-gallon hat against his chest as he talked with a lady.

Nobody went hungry that day, I am sure. It was a pleasant sight to see families seated near their wagons and eating from well-stocked baskets. Home-grown chicken fried to a golden brown, eaten with fresh-baked bread spread with buffalo berry jam and with cold tea or milk as a beverage, makes a satisfying repast when eaten on the prairie in the bright sunshine of a September afternoon.

My horse, Tony, was treated to hay and oats at the livery stable. Then I sought out the tent where the church ladies were serving meals at the modest price of twenty-five cents. Chicken was there in abundance, and heaped plates of baking powder biscuits were replenished as necessary. The women of that frontier were adept cooks; biscuit and cake mixes were undreamed of at the time.

Cooking was done on a range stoked with stove wood: it took an expert to regulate the heat. Baking loaf bread and pies was the real test. I remember the apple pie served that day—it was delicious!

The rodeo staged that afternoon on a field west of town provided plenty of thrills to astonish the onlookers, many of whom, including myself, had never seen a rider atop a bucking horse. The contestants that afternoon were not professional riders, but many of the boys gave superb performances, fanning the air with their hats, and holding the reins clear of the saddle horn. I held my breath when one was unseated to land on the ground in a cloud of dust, but soon found that this was one of the expected bits of excitement without which no bucking contest would be complete. There was no big money for the successful riders; most of them took part for the fun of it, so I was told.

As I rode home in the late afternoon wisps of high cirrus clouds drifting into the sky picture indicated that a change in the weather might be in the making—something that could be expected after the Indian summer days which we had been having. Sure enough, next morning we looked out of our cabin home at Muir to see a land blanketed with new snow. Angry looking clouds over the Little Belts gave promise of more to come. Far ranges—the Crazies and Snowies—had their heads in the clouds. After a late breakfast I saddled Tony and rode up Hopley Canyon which had all the beauty of a winter wilderness. Tracks of little wild creatures, and big ones as well, were everywhere in the new-fallen blanket. Clark nutcrackers were busy in the limber pines, their black, gray, and white plumage conspicuous among the branches where they were extracting seeds from the cones. This was the first time I had seen the bird to identify it and the first chance to become acquainted with the characteristic raucous call, not quite the same as the note of any other jay. Ornithologists recall that the type specimen of the bird was collected by the Lewis and Clark party; the species was named for Capt. Clark.

Soon we received instructions to discontinue the survey project. Lou and Jim were to return to Neihart; Sutton and I were to go to a tree-planting project at Dry Coulee in the Little Snowies Division of the forest. Sutton had business which had to be attended to before he joined the planting crew, so he turned his horse, Sailor, equipped with the ancient pack saddle, over to me, and took off across the hills to Neihart, carrying with him my promise to pay for my new steed come payday.

It was a long ride to Dry Coulee. But the weather had turned warm, and the snow disappeared rapidly. The sun, shining from unclouded skies, bathed the benchlands with a mellow warmth; one found it comfortable to ride with coat tied behind the saddle and sleeves rolled up. Gleaming mountain ranges drew close in the clear, washed air. Aspen groves among distant foothills stood out in orange dress to remind one that frosty nights had been there, and more would come. Pausing to eat lunch in the shade of a strawstack, I detected an unguessed chill in the air; the sunshine had lost the fiery punch of mid-summer.

Dry Coulee Ranger Station had come into prominence because of a reforestation project initiated that fall. The customary practice in those days was to have stations built within the boundaries of the national forest, and Dry Coulee was no exception. The ranger in charge at that time was Clyde Knouf, who had recently arrived with his family, consisting of a wife, son and daughter, from Cass Lake, Minnesota. At the time of my arrival Clyde's chief worry was not concerned with tree-planting, but where and how his youngsters would go to school. The remoteness of the Dry Coulee station from even a satisfactory rural school posed a real problem, which I think was solved at last by Mrs. Knouf—who established a home in Harlowton for the school year. Al Rossman had also come from Minnesota with his family at the time the Knoufs made the change. Both Knouf and Rossman were experienced log scalers, and were later transferred to the timbered part of the region, where real use of their talents could be made. Knouf became one of the top scalers of the entire Forest Service.

The fire-swept benches on each side of Dry Coulee were chosen as planting sites. Only a small percentage of the area of old burn showed signs of natural reforestation, and these benches apparently were ideal for an experiment in artificial planting. At that time we had a lot to learn about where trees would grow, the best methods of planting, the most suitable species for different locations, spring versus autumn planting, the proper method of shipping stock, and caring for it when it was received from the nursery. The Savenac Nursery at Haugan, in extreme western Montana, had been recently established, and it was there that our little trees were obtained. These were three year olds who had spent one year in the beds where they were started and two more in beds to which they were transplanted to prevent crowding as they developed strong roots. As near as I can recall, ponderosa pine was the only species used on this particular project.

The bundled trees were shipped by railway express to a station on the Milwaukee Road, which I think was Lavina, and picked up there by a horse-drawn wagon for the long ride to Dry Coulee. Today the round trip would take about two hours; then it meant a long hard day—hard on the driver as well as the team. Once at the ranger station the bundles were opened and the bunches of a hundred trees each would be placed in trenches and the roots carefully protected with moist earth. This operation went by the name of "heeling in," and was considered one of the most important steps in the planting process.

Planting in the field was done by two men working together. One man would make a good-sized slit in the ground with a mattock; his partner would place the seedling in the hole with the roots spread out, fill the opening with fine dirt, and tamp it down firmly. Six feet further on the same process would be repeated. Rows were spaced six feet apart. A two-man team would plant from 900 to 1,200 trees in an eight-hour day. Possibly two dozen men comprised the whole planting crew. Many of these men were from local homesteads, and a fine bunch they were.

One of my jobs was to superintend the heeling-in of the young trees. The beds where this was done were in an aspen grove, which during our stay in Dry Coulee turned to a brilliant orange. The interior of this grove on a sunny day was flooded with golden light, and tempted one to linger and enjoy not only the color but the shade which tempered the bright October sun. On more than one occasion I ate my sack lunch there, breathing in the odor of the moist soil under the aspens.

R.Y. Stuart, then chief of the office of silviculture, came over from Missoula to inspect the work, and I guess to check up on the new forest assistant. It was a delightful evening that he spent with us, telling of the beginning days of the newly formed Forest Service and what it stood for. Stuart was a dedicated man, and had the ability of firing others with some of his zeal. Although we may not have realized it at the time, that evening gave our little group -rangers, part-time men, and a forest assistant—a new understanding of forest conservation and a renewed incentive to give the work the best we had. It was quite a trip over from Missoula in those days, but in this case the journey paid off. Incidentally, Stuart taught me to identify the limber pine!

My horse, Sailor, with his ancient packsaddle became indispensable once the job was underway, and it was not unusual for me to make the trip from the heeling-in beds to the planting site several times each day with a load of trees. The seedlings were carried in two special pack boxes hung on each side of the saddle. Keeping the men supplied with planting stock by this primitive means of transportation was quite a task. Looking out from the trail that climbed the bench, one had fine views—south to the Musselshell Valley, north to the nearby Snowies margined by the shelving ledges known as the Washboards. Looking forward to a Sunday, I determined to climb to these unique formations; this I did on an afternoon when the weather could still be classed as Indian summer. There wasn't much of a trail, and after the canyon became more rugged I tied my horse, Tony, to a small limber pine and scrambled up to the Washboards on foot. The rock appeared to be limestone, and in places was filled with fossil shells. Many of these fossils were scattered about and could be picked up by handfuls. The rugged character of the terrain and the unique character of the sloping limestone ledges gave the spot a singular wildness. I rested a while on the bare rock which had been warmed by the October sun, gathered a few fossils, and then descended to the spot where I had left my horse. According to the map, I had been almost as far as the head of Horsethief Canyon.

One day a cold storm came out of the north and buried the Coulee in a lasting mantle of snow. We knew that the planting season was over. The crew stayed with the job until the few seedlings remaining to be planted were in the ground and then left—some for nearby homesteads, some to join the tide of migrant labor drifting to warmer climes. In a few days the camp was deserted except for a trustee or two who picked up tools and tidied the premises.

My orders were to return to Neihart with my horses. This would be a ride of several days, and I looked forward to it with pleasant anticipation. Could I have foreseen some of the rigors which lay ahead, I might have taken a less jubilant attitude. Looking back, however, I see that it was an adventure which I wouldn't have missed.

The route was outlined for me by Henry Sutton, and his directions were quite specific. I was to head for the Judith Ranger Station, spend the night there, and then follow up Yogo Gulch to the divide from where I would follow down Dry Wolf Creek to the Wolf Creek Ranger Station for another night. From there one more day would suffice to reach Neihart. "Not much snow in the hills yet—you won't have a bit of trouble," said Henry.

This was the first part of November, and the days were short. It was nearly dark when I rode up to the Lemmon Ranch near Garneill and put my horses in the corral. I was made welcome to the bunk house, which was comfortably warm after the ride through the crisp autumn air. It was an ample meal set before the hands and guests that evening. No one talked very much around the table, but the atmosphere was jovial; the light from the single large kerosene lamp fell on faces that were friendly. Some of the womenfolk hovered about the stove keeping watch over enormous pans of baking powder biscuits and dishing up the famous stew known as mulligan. I did not go to bed with an empty stomach that night!

The ranch day began early, and breakfast was eaten by the same kerosene light. Clouds had covered the sky during the night, and daylight did not come until I had been on the road an hour. B.F. White's ranch was reached for lunch. and I spent the night with the Phillips outfit well over towards the Judith River. Next day before noon I reached the Middle Fork Ranger Station and met for the first time Guy Meyers and his recent bride. I became very fond of these fine people in the ensuing years, and followed Guy's successful career in the Forest Service with much satisfaction. At one time he was superintendent of the Ninemile Remount Depot.

Guy was a large man, rarely without a smile. I remember the pleasant voice with which he greeted me as I rode up to the station and dismounted. Almost immediately Mrs. Meyers came out of the house to join in the welcome and I suspect to size up the new forest assistant. She had been a well-known and well-liked schoolteacher before she married Guy. Of fine bearing and well-educated, she took her place in the Forest Service family with credit.

The next day was Sunday, and no effort was made for an early start for the trip up Yogo Gulch and over into Dry Wolf Canyon. It had snowed in the night, and the prospect was wintry as I started about ten o'clock. A bitter wind was blowing out of the north and gave promise of more storm to come. Frost formed around the noses of the horses, which, coupled with the endeavor to keep their heads down out of the wind, gave them a truly dejected appearance. Ice crystals sifted under my coat collar and pull-down cap and my feet and fingers became numb from the cold. Soon the diamond hitch which held the bedroll and duffel on the saddle had to be adjusted, which with my limited experience proved to be quite a task in the chilly air.

At the mention of Yogo Gulch one familiar with the locality thinks immediately perhaps of the sapphire mines. The workings, one English and one American-owned, are interesting as examples of a unique mining activity. Years after my initial ride up the gulch I paid another visit to the country in more clement weather. On the first trip blinding snow obscured much of the landscape and the mine buildings, and there was no temptation to stop for sightseeing, but on this later trip I found the air delightfully warm, with shadows from great cumulus clouds chasing each other across mountains and foothills. Limber pines, whose picturesque growth tempts one to linger for pictures, add much to the scene. It is a charming country—in good weather! Then too, the garnering of the sapphires is an interesting process to investigate.

On that afternoon no pictures were taken, except perhaps mental impressions which have persisted to keep alive the memory of that stormy passage of Yogo Gulch—the horizontal surges of stinging sleet, the swaying tops of the trees, log cabins along the trail, the indistinct bulk of the Wetherwax smelter looming large through clouds of blinding snow.

Night was falling when I climbed towards the divide at the head of Yogo, and a dreary scene it was. Deep snow was on the ground, and this was being whipped into drifts which came to the stirrups in places. The horses slipped and floundered, and it was hard to keep them faced into the gale. No semblance of a trail was visible, and the telephone line which we had been following had disappeared. The chances of reaching Dry Wolf Station without an all-night ride seemed remote indeed. It was long since breakfast, the last meal I had eaten. And then came the realization that I had brought no food along. Matches and bedding, yes, but meals, no. Turning the animals around, I back-tracked to the old smelter I had passed some miles down the gulch. It would do for a night's lodging. Any port in a storm!

The building was a huge log structure, long since dismantled of machinery and with interior scaffolding in a state of disrepair. It was said that Wetherwax, the builder, had been killed by a fall from this same scaffolding. It was a somber spot in the twilight of the winter day, details of the interior barely discernible except for white drifts in the corners and a snow cover over most of the floor. With every gust puffs of snow would be forced between the chinks of the logs and dusted over objects scattered about in the gloomy, cavern-like room. But I was grateful for the shelter the old building afforded, and so were the horses. On the earthen floor I built a small fire of wood salvaged from crates, and prepared to go supperless to bed. Canvas and sugans were spread on the ground not far from the horses, who stood patiently nearby with heads down, relieved perhaps to be sheltered from the gale which was raging outside. Lying snug beneath my quilts I listened to the blasts as they seemed to merge into a continuous roar; whenever I poked my nose out from the warm bedroll a fine mist of ice crystals stung my face. But the dynamic fury of the storm seemed to bring more a feeling of exhilaration than of positive hardship, even without supper, and I well remember how fortunate I felt to be having this exciting experience as part of my official duties. The fire died down at last, fanned to extinction by spurts of wind, and tired and hungry I fell asleep.

The storm had cleared somewhat in the morning, but it was bitterly cold. I was hungry, and so were the horses. All signs of a trail were obliterated by drifts in the gulch, and no telephone line was visible. Not knowing the country and being desperately hungry, I felt the wisest way out of the predicament was to go back to the ranger station, feed the horses and myself, and get a fresh start. This I did.

I shall never forget the look on Meyers' face when I rode into the yard where he was splitting wood. He assured me I had done the right thing in turning back, and admitted it would have been better for me to have delayed my start of the day before until the storm had cleared. By comparing notes it appeared I had missed the trail in the gulch, which had been buried in snowdrifts, and had failed to notice the telephone line which could have been relied on to lead me to the crest of the ridge.

Food never tasted better than the lunch that Mrs. Meyers set before me on the table in the cozy kitchen. That finished, I spent the remainder of the day helping Guy split and pile wood, an agreeable task in the bright winter sunshine which had replaced the bitter weather of the day before. The horses were fed oats and hay in the ranger station barn, Tony giving a whinny as often as I came near the door. The night in the smelter seemed more of a dream than an actual experience.

An early start was made the next day; by noon I had reached the Dry Wolf divide and looked over a mountain world glittering white under a cover of new snow. The drifts had combed over on the crest of the ridge to form an escarpment up which the horses had to struggle. Sailor lost his footing and floundered around in a most agonizing way before he could regain his feet. Tony did better, and carried me to the top. On the ridgetop the rocks and clumps of beargrass had been swept clear of snow, giv-

*Splendid Was the Trail*

ing a more stable footing and a chance for the panting animals to catch their breath. I relaxed in the saddle to drink in a panorama still vivid in memory after a lapse of over fifty years. At our backs was Yogo Gulch up which we had just come; ahead and below lay Dry Wolf down which we would soon be heading. Everywhere mountains were in the picture—the Little Belts, the Snowies, and far to the northwest the Highwoods. All the country was dazzling white. I had no snow glasses to cut the glare but squinted bravely to drink in the glorious sight.

Ray Evelith was the ranger in charge of the Dry Wolf district. Meyers had called him on the telephone, telling of my departure from his station, and asking him to call back as soon as I arrived. This was done, which I am sure brought a sigh of relief to the folks over on the Judith.

The next morning I was on the trail early for the last leg of the journey. It was a tedious struggle through deep snow to the divide where we again stood on a knife edge blown clean of snow. From there it was all downhill to Neihart through a snow-filled lodgepole forest where we made the first horse tracks since the storm. That afternoon it turned bitterly cold, and when we rode into the little town where lights were beginning to shine from the windows, the thermometer stood well below zero. I had the feeling that the long ride from the Snowies was finished in the nick of time. Hay and oats for the horses and food for a hungry rider would both be forthcoming. The smell of wood smoke drifting from chimneys and stovepipes gave promise of ample protection from winter cold in the days ahead. Neihart seemed home to this weary traveler.

Life in a snowbound village proved to be much more interesting than one would at first suppose. Not many people were working there at the time—a few miners, several men engaged in cutting firewood and stulls, one or two hauling and shipping baled hay from the high parks. On week-days the arrival of the train which brought the mail was looked forward to with pleasant anticipation. The post office, in charge of a family named Larson, was a gathering place at mail time, and there one heard the latest local news, and shall I say it, "gossip"? But a spirit of friendliness prevailed in the tiny office crowded with mail-seekers, and one seldom heard unkind remarks voiced about anyone or anything.

Mailtime for me was of particular significance, for it often brought of-ficial letters of instruction from the office in Great Falls, bulletins and re-ports to be read and returned, and calls for information which I was in a position to furnish. Well do I remember tearing open one such official missive in the stuffy little post office and reading, "You will examine and report on stands of timber suitable for poles in areas tributary to Neihart, submitting maps showing location of same, and giving rough estimates of volume." Here was a real challenge. Nothing like it had come my way be-fore, and I may say, never has since.

This was mid-December. Snow was already so deep that only on snow-shoes could one travel through the timber. Extensive stands of lodgepole suitable for power poles were rather remote from town and at least a thousand feet higher. However, with the exuberance of a young and inexperienced forest assistant I tackled this job of winter reconnaissance. Snowshoes were hard to come by, but I was fortunate in getting the loan of a pair from a miner living in town. A team was still hauling baled hay from the Ledbetter ranch in Island Park where there was a log house used by the hay hands in summer. The driver of the team agreed to haul my outfit to the ranch on the hay sled, and to keep in touch with me on his daily trips; I was given permission gladly to make my headquarters in the house.

Deep sled tracks were well-filled with drifting snow as we headed up the valley, and it took us most of the forenoon to reach Island Park. Clapping his chilly mittened hands together, the driver remarked that probably this would be the last trip of the season as far as he was concerned. There was just an unbroken expanse of deep snow between the point where the sled turned around for the return trip and the house, which was some 300 yards away, and I had to break trail with snowshoes and backpack my belongings over this final leg of the trip. Long before I had finished this chore, the sled had been loaded with hay and was homeward bound, leaving me absolutely alone in what seemed a vast wintry world.

The old log house was poorly chinked and draughty. The small kitchen stove was totally inadequate to warm the frosty, dismal room, and I soon found that only by sitting close to the open oven door could any semblance of comfort be obtained. The little monster consumed enormous quantities of dry lodgepole which I brought in from the woods which surrounded the park. I soon had a well-packed snowshoe trail from the kitchen door to the woodlot, a distance of two hundred yards or more. Small tree trunks were cut into stove lengths with an old-fashioned buck saw which I found on the premises. It was anything but a gay life, eating meals kept hot on the stove top, melting snow for water in a pail improvised from a five-gallon oil can, and sleeping on the top of a table drawn close to the fire. A couple of drapes were pulled from the windows to provide more bedding. The evenings seemed very long; by the feeble light from a kerosene lamp I washed the day's dishes and thumbed over notes taken in the field. Bed seemed the most comfortable spot in that dreary world.

With one exception it snowed on every one of the five nights I spent at the ranch. On that one evening I poked my nose outside to find the world glittering white under a full moon. Clothed against the cold I put on my snowshoes and walked over to the woods. The air was perfectly still, but every once in a while a snow-laden branch would release itself from its load, giving rise to a small cloud of silvery dust. Flecks of moonlight dappled the shadowy forest floor. Strangely enough it seemed a friendly place, the lodgepole forest on that cold winter night—a good place to be despite the sub-zero temperature. It was with some reluctance that I left the scene to return to my billet beside the over-worked little stove. One thing more clings to memory—the lonely hoot of a great horned owl, telling me that other life was abroad in that cold world.

Finally on a snowy morning almost a week from my arrival at the

*Splendid Was the Trail*

ranch, I buckled on my snowshoes and headed for Neihart by the shortest practicable route. My bedroll was left behind to await the return of the hay team; when I would recover it was a problem. My trail led through a large stand of over-mature timber north of Island Park and then down the steep slope to O'Brien Creek a mile or so above the dam.

Then Neihart! No metropolis could have seemed more welcome than the ramshackle little town strung along the hanks of Belt Creek. The friendly human voices which greeted me at Mrs. Mix's boarding house were sweet indeed!

A week later the hay team re-opened the road to the ranch and brought out my bedroll. My notes were put in order and forwarded with a map to the supervisor's office. What use they were put to I never learned, but I presume the timber which I cruised has long since gone to the poleyard and mill. Still with me, however, are the memories of the lodgepole forest in winter—the tapering trunks veiled by failing snowflakes, the crunch, crunch of the snowshoes on the white snow cushion the only sound to break the silence.

Well do I remember a cold, blustery day in the latter part of January when the mountains were obscured by dark clouds and horses at hitchracks tried to stand with their backs to the wind. On the way to the post office I fell in with several townspeople who remarked we were in for more snow. The mail was up and folks were going away with the usual assortment of letters, newspapers, Sears Roebuck catalogs, and packages both large and small. There was only one piece of mail in my box—a letter in an official franked envelope. Opening it I read, "You will report to Missoula as soon as possible to assist Frank Rockwell in the preparation of yield table." This was to be my good-bye to Neihart, which I did not visit again for many years.

# Pine Hills and Rimrocks

My work with Rockwell in Missoula might be summed up bluntly as an attempt to bring order out of chaos. A vast amount of data for the preparation of yield tables had been sent in from the field, and this material had to be assembled and tabulated. I was the Man Friday. My superior was considered one of the ablest research men in the Forest Service, and I considered it a privilege to work with him and observe his patience and meticulous attention to detail.

*Splendid Was the Trail*

I never lost the high regard I had for him. But feeling as I did that the pleasures of field work were more to my liking, it was somewhat of a relief when I was told by John F. Preston of the District Office that I was being assigned to the Sioux National Forest in extreme eastern Montana and western South Dakota. "Your headquarters will be in Camp Crook, South Dakota, a town still on the frontier, two day's ride from the nearest railroad," commented Preston as I left his office.

At that time the Dakota Forest, an area on the Little Missouri River in North Dakota, was administered from Camp Crook. Here experimental planting work was being done, the young trees being supplied from a small nursery maintained under supervision of the ranger. My duties would include periodic trips there to assist the man in charge with the nursery and planting projects.

Supervisor Ballinger, a kindly and most considerate man, had left his horse in a livery barn in Bowman with the request that I pay a visit to the Dakota Forest before going to Camp Crook; that would mean a ride of approximately 200 miles before I turned the animal over to his owner. Automobiles were not in general use for prairie travel at that time; the horse was considered a most dependable means of transportation, and a fifty-mile ride in a day was merely a matter of routine.

Spring had touched the prairies as I rode north from Bowman on a bright April morning. There was a shimmer of green on the rolling hills, and shrubs and trees showed signs of leafing out along the coulees. It was an era of homesteading. Many fields had been fenced and planted to grain or flax. I noticed the homes that these newcomers had built—some of them made of sod, others frame shacks covered with tarpaper. I soon learned to detect the pungent smell of burning lignite coal, a fuel which did much to make the settlement of this prairie country possible. I remember seeing a homesteader digging coal from a bank near a coulee bottom. We had a little talk before I rode on.

Before leaving the Regional Office in Missoula (then called the District Office), I was given rather complete instructions as to what my duties would be as Forest Assistant on the Sioux and Dakota Forests. I was to become thoroughly acquainted with the far-flung divisions of the Sioux— the Long Pines, the Short Pines, the Ekalaka, the Cave Hills, and the Slim Buttes, for the purpose of preparing a silvical report and a plan for the best use of their resources. My duties on the Dakota Forest would deal primarily with planting projects. I was to assist the ranger in expanding the forest nursery and selecting suitable sites for planting the young trees. As before stated I had gained considerable experience the previous fall on planting projects in the Big Snowies of central Montana, and R.Y. Stuart felt that this might prove helpful in getting a planting program under way on the Dakota and Sioux Forests.

The Logging Camp Ranger Station of the Dakota Forest was on Deep Creek, a tributary of the Little Missouri River. One saw it first from the head of a broad swale which led down from the higher prairie. It seemed an oasis among scoria buttes and badland bluffs on which were growing scattered ponderosa pines and junipers. There was a long one-room building, with a screened-in porch the entire length of the south side, and a substantial gambrel roof barn across the yard at the rear. The small nursery was near the creek south and west of the buildings. This nursery was irrigated by water pumped from the creek, which, as I remember, never ran dry. Green ash and other small trees grew along the stream and provided welcome shade on hot days.

Ralph Sheriff was ranger in charge at the station. My first meeting with him came as I entered the building and found him taking a siesta on the bed surrounded by several cats. Ralph was a graduate of the University of Illinois and had come to the western Dakotas with one of his college chums named Haines. The boys had worked for some time building sod houses for homesteaders in the country around Lemmon and Hettinger, and then decided to take the Civil Service examination for the position of Forest Ranger in the United States Forest Service. Both passed. Haines was appointed to help Supervisor Ballinger in the Camp Crook office of the Sioux; Sheriff got the job on the Dakota.

Sheriff was a very capable man—practical and able to do many things well. He was good at handling men and was well liked by all who worked for him. In dress he was quite unconventional, but he was a man of cleanly habits. He was always smiling and nothing ever disturbed him much—a good quality for a forest officer in those days. He was of medium height and rather stocky.

Work in the nursery was in full swing shortly after my arrival. Seedlings of ponderosa pine had to be transplanted from the beds where they had been grown from seed to other beds where they would be evenly spaced and have room to develop until they were three or four years old and ready to be set out in the field. A device known as the Yale planting board was used for transplanting. This consisted of a narrow board with notches in which the seedlings were placed so that the roots extended beyond the edge of the board. Another board of corresponding size was hinged so that it could be closed down on the crowns of the seedlings. When lifted, the evenly spaced seedlings were held securely in place with their roots hanging down so that they could be placed in a trench made ready for them. After the earth was firmed around them, the planting board was opened and removed for another loading of seedlings. The roots of coniferous species are very easily damaged by exposure to drying wind or sunlight and must be kept in the shade during transplanting operations. For this purpose a rough booth was constructed of canvas tarpaulins.

Homesteaders, known locally as "honyocks" (meaning of term obscure), jumped at the chance to pick up some badly needed cash by working in the nursery. There were several young couples whom I remember well, although I cannot recall all their names. One tall boy, Harry Roberts, was at the time courting the girl he afterwards married. I also remember

*Splendid Was the Trail*

Joe Miller and his sister Marjorie, who later became Mrs. Ralph Sheriff. Several older folks also took part.

Travel home at night for most of these people was impossible, so they camped at the nursery. I remember many happy evenings spent around the campfire exchanging stories or listening to music played on the violin and guitar by two of the talented persons of our little group.

After transplanting was finished, considerable field planting was done on various areas in the vicinity of the ranger station. I believe one of these areas was on Sand Creek. Planting was done by two-man crews. One man would dig a hole with a mattock; the other would place the seedling and press the dirt firmly around the roots. Many of the seedlings were set in the loose soil on the slopes of the scoria buttes. Rattlesnakes were a menace, and one had to be on constant alert when planting in these locations.

Much of the stock set out in these operations came from the Savenac Nursery in the Lolo Forest of western Montana. It was shipped by rail to Bowman in bales protected by burlap and transported to the ranger station by wagon. It is believed that a good deal of this stock was from seed collected in the Black Hills (Pinus ponderosa, var. scopulorum). Eventually, young trees from the Dakota nursery would be used for planting, but at this time no stock of the right age was available from this source.

We felt at the time that the best planting sites were on slopes where some tree growth was already established rather than on areas which were more or less flat and where the seedlings would have strong competition from the prairie grasses. Where planting was done on grassy land the trees were set in furrows made by a sod-breaking plow. In the more rolling terrain where pines of considerable age and size were growing were sites which seemed well adapted for successful planting. There were also north slopes which were partially protected from the hot sun and also from the drying winds that swept across the prairie from the south and west. The soil in these locations was more or less loose and seemed capable of soaking up moisture readily. Here, in contrast to the heavily sodded areas, there would be much less grass competition. Whether or not our surmises were correct, I do not know. Studies were never carried to completion, to my knowledge.

Although the time I spent on the Dakota National Forest totaled no more than a few weeks, I grew to love this part of the Little Missouri valley and could well understand the fascination the region held for Theodore Roosevelt when in the eighteen eighties he ranched in the vicinity. Near the ranger station there was a high point overlooking the river where I often rode of an evening. Here the river made a great bend, and across the valley rose the Tepee Buttes-I particularly recall going there one hot evening when the moon was full. That night I had for company Ethel Sheriff, Ralph's sister, an Illinois school teacher out "West" for a visit. Dismounting to enjoy the view we sat down. There grows in that part of the country a certain species of prickly pear cactus. Oldtimers are well acquainted with it. My companion was not. She did, however, overcome this deficiency immediately!

On Sundays Sheriff would often hitch up the team and we would go on exploring trips to far parts of the forest. Perhaps the most interesting feature we visited was a burning coal mine. Here a bed of lignite coal was burning below the surface of the ground and through fissures we could see the subterranean fire which had already eaten away the fuel under a considerable area, causing the ground to slump. How long this coal had been burning, or what touched off the fire was a matter of speculation. The best guess seemed to be that lightning was the cause, but how long ago the strike came nobody that I talked with knew. That the red scoria buttes throughout the badlands are the result of prehistoric fires which cooked overlying beds of clay into a natural brick seems to be the opinion of the geologists.

Growing in the vicinity of this burning coal mine are many slender juniper trees which attract attention by their beautiful form. Neither Sheriff nor I could account for these trees, but in 1946 when I made another trip to the locality I got the explanation from William Hanson, who at the time was majoring in Botany at the University of Montana. He told me that, according to Professor O.A. Stevens of the North Dakota State College at Fargo, they are a hybrid, Juniperus scopulorum columnaris. The original study and classification was made by Dr. Fassett, a leading taxonomist at the University of Wisconsin.

Sheriff and I often called at the Hanson Ranch, or Logging Camp Ranch, a beautiful spread in the Little Missouri bottoms. At this point ties which had been cut for the Northern Pacific Railroad, then building, were dumped in the river to be floated down to Medora. I believe many of these ties must have been cut in the Long Pine Hills. There is a Tie Creek rising in these hills just west of Camp Crook. Possibly some of the ties were also cut from the breaks and swales near the ranch. The operation was not successful, as the ties got hung up in the shallows along the river and the cost of salvage was prohibitive. The name Logging Camp is all that remains as a reminder of this episode.

Other bits of history connected with the Logging Camp Ranch are interesting. We were told that the last mountain sheep in North Dakota was shot from the yard. It was standing on a butte in plain sight from the house. The house itself had been hauled overland in two sections from Dickinson by a man of Russian extraction. The move was costly and the man never recovered from the financial loss, so it is said.

The Hansons were good neighbors and always ready to lend a hand where help was needed. Western hospitality in its best tradition reigned at the ranch. To me, these friends exemplified the finest spirit of the true Western pioneer, and contact with them went a long way in helping me, a city-bred boy, to become oriented to a new way of life.

My first visit to the Dakota ended with the conclusion of the spring planting season early in June. On my way to Camp Crook I stopped for a few days in the Cave Hills, meeting the ranger, Dave McGill, and talking over plans for making a plane table map of his district later in the summer.

*Splendid Was the Trail*

Camp Crook was a remote spot in those days when roads were little more than rutted tracks across the prairie. It took two days of horse travel from Bowman to reach this tiny hamlet on the Little Missouri River some eighty miles to the south. Shortly after my arrival there a motorized stage was put into service to bring mail from Bellefourche on the northern edge of the Black Hills. In fine weather this trip took less than half a day; after heavy rains wheeled vehicles would become mired in gumbo mud and traffic might be halted for a week.

Forest headquarters were housed in a one-story frame building on the south edge of town. A boldly lettered sign on the front read "Sioux National Forest." The front room. which served as an office, was furnished with bentwood chairs, filing cabinets, an Oliver typewriter, and other rudimentary equipment necessary for the transaction of forest business in those beginning years of the Service. A large stove which burned enormous quantities of lignite coal and produced corresponding amounts of ash stood in one corner. One of two rooms in the back served as storage space for closed files, fire-fighting tools and the like; the other as private sleeping quarters for the supervisor, Charles Ballinger.

The main street of Camp Crook, along both sides of which straggled the mercantile buildings, the hotel, the bank, the saddlery, the saloon, was dusty in dry weather and a quagmire in wet. Plank sidewalks with many ups and down enabled those on foot to avoid the mud to some extent, provided they did not have to cross from one side of the street to the other. Conspicuous in the very middle of this street was a bandstand. There was no lack of hitching rails for horses, and when trading was brisk it was pleasant to saunter and note the many saddle animals and teams tied up while their owners transacted business—perhaps in Chuning's Mercantile, the bank, or with old cronies at the saloon. Crook might be isolated, and the main street rutted and unkempt, but the wind blew sweet off the prairie, and occasionally brought a whiff of pine-scented air from the timbered hills to the west. One felt that Nature had been kind to the little town, and had breathed vitality and a spirit of mutual helpfulness into the inhabitants. It was often said in those days that having once lived in Crook one always wished to return.

The Sioux National Forest, as I found it, was a far-flung empire. Included were the Chalk Buttes, the Ekalaka Hills, the Long Pine Hills, all in extreme eastern Montana. In South Dakota not far south of Camp Crook were the Short Pine Hills which included much state-owned land. Farther east were the Slim Buttes and the Cave Hills.

Forage values of these scattered units were fairly well known, but little study had been made of their timber-producing potentialities. How fast were trees growing on the varying sites of these buttes and rolling hills? It was my mission to find out. In addition to preparing a complete silvical report of the forest, I was instructed to compile a series of yield tables based on detailed growth studies.

Alex MacNab, ranger in charge of the Long Pines division, kept bachelor quarters at the Tie Creek ranger station just over the state line in Montana a few miles west of town. A slight, soft-spoken man with a Scotch accent, always scrupulously neat in appearance, Mac was well liked by all with whom he came in contact either officially or socially. I count myself fortunate to have had Mac as guide and counselor when I was inexperienced in the ways of a country which was new to me.

Mac's favorite mount was a thoroughly spoiled bay horse named Whistler. Old Whistler was essentially gentle, but he was strictly a one-man animal, seldom, if ever, ridden by anyone but his owner. Even Mac of a cold morning would sometimes have difficulty in staying aboard. Pending the purchase of a horse of my own I had been riding an animal lent to me by Supervisor Ballinger. It was a gentle horse of which I became very fond. It was he who carried me safely across the Little Missouri in flood-the first and last time I have ridden a swimming horse. One morning it appeared that Ballinger needed this horse for his own use, and Mac suggested I borrow one of his which was then in pasture at Tie Creek.

"Ride out to the station and get saddled up," said Mac. "I'll be along shortly, and after lunch we can take off for an over-night trip around the forest." Then he gave me a description of the animal I was to ride. "Get some oats and he'll come right up to you; he's plumb gentle," he added.

Yes, he came right up to me and nuzzled the oats, as Mac had said. He was beautiful and sleek, and surprisingly easy to saddle. A few more munches of oats and I was on top-but not for long! The bucks were long and graceful, and in perfect rhythm, but they sufficed to land me in the soft muck of the corral. By mistake I had saddled Whistler!

Riding across the rolling mesa top of the Long Pines that afternoon I got my first close look at the characteristic forest growth of these timbered hills. It was charming country, with patches of trees interspersed with grassy parks. In some spots there were cultivated fields. Ponderosa pine was predominant, and the only species of much commercial importance. On south slopes and along the rimrocks exposed to the hot sun and drying winds growth was extremely slow, but on north slopes and in the moist gulches trees grew to considerable size, and stands of old growth timber provided logs for several small mills. The lumber, although not of the highest quality, was of inestimable value to ranchers, homesteaders, and townsfolk living in this country so remote from other sources of supply. Other products there were to meet the needs of prairie dwellers who converged on these hills from all directions to carry home fence posts, corral poles, or firewood. Resinous pine was highly prized for kindling. Even dead limbs were sometimes salvaged from standing trees, and knots and branches gathered up carefully from the ground. Dead timber was free for the asking. Much the same characteristics and conditions prevailed in the Ekalaka Hills and on the other divisions of the forest.

It was June, and this afternoon was windless and hot; the sun beat down from a cloudless sky, bringing a feeling of drowsiness which left us completely relaxed in the saddle. The horses, too, felt the heat and slowed their pace accordingly. At last we left the table land and dropped into a timbered gulch which led down to Whitcomb's mill. Here tall pines, standing fragrant in the breathless air, cast shade which gave some relief from the heat. Below we could hear the mill noises, the whine and growl of the saw, the thud of the logs as they were rolled onto the carriage.

This mill was perhaps the best of those operating in the locality at that time. The one circular saw cut about 15,000 feet of lumber in an average day's run. The owner, Ralph Whitcomb, was in my opinion a progressive operator. His mill had a tidy look, and I always found him fair in his dealings with the Forest Service.

Both Ralph and his wife were genial people and extremely hospitable. Mrs. Whitcomb was a good cook, and forest officers often planned their trips so as to be there at mealtime. It was understood that meals would be paid for at the going rate of fifty cents, and this rule was always observed on official trips.

There were logs to be scaled that afternoon, and Mac gave me my first lesson in the use of the Decimal C scale stick. The logs were piled loosely, and I remember well how difficult it was to get a good look at individual pieces in order to find defects for which allowance was to be made. Mac had scaled for many years, and I found him an able teacher. Under his tutelage I eventually became proficient enough to be given from time to time the job of scaling at the Jennings mill, which was in the northern part of the Pine Hills.

There is little scenery in the Long Pine and Ekalaka Hills which could be classed as spectacular. But there is a distinctive charm to vistas of plains and broad valleys framed by venerable trees growing along the rimrocks—drought resistant veterans centuries old, whose taproots have found sources of moisture deep in the ground. These sturdy trees, often stunted and fantastic in growth, are pioneers that can be lived with and loved. Many were growing on these escarpments when herds of buffalo surged across the prairie, pursued possibly by Sioux hunting parties.

On our return from Whitcomb's mill, I had my first look at Bell Tower, an eminence of white stone at the southwest extremity of the Hills. It rises some hundred feet or more from its base, and from the summit one has a view across the valley of Box Elder Creek to the Ekalaka Hills and the Chalk Buttes. It proved to be a fascinating spot, made a little awesome by Mac's stories of a rattlesnake den supposed to be in a crevice of the big rock—a place of hibernation for countless snakes which spread far and wide with the advent of warm weather. It was said that cowhands and sheepherders often dropped rocks into the den to stir up the reptiles and make them "sing." This might have been more than a tall story. Years later South Dakota employed a snake exterminator who worked part time in this locality. In one of his traps I once saw 88 snakes, mostly rattlers. This was on Battle Ridge, not far from Crook.

Forest officers had many contacts at that time with homesteaders both on and off the forest. MacNab was universally liked and respected by these people. We didn't use the term "public relations" in those days - maybe it hadn't been coined at the time, but Mac was a true exponent of the expression in its best sense. Never a shack or soddy where this genial ranger with his Scotch accent was not welcome. On his part Mac was a sort of "who's who" of the cooks of the locality, for he knew well where the finest biscuits or the most delicious dried apple pies were baked. Itineraries were planned accordingly. On more than one occasion Mac and I cooked a meal during the absence of the householder, washed the dishes, and left the price of the meal on the table. Door keys were seldom, if ever, used in those days!

On my return to Camp Crook after this initial trip with MacNab I bought a saddle horse—a beautiful sorrel named "Kid," for which I paid $100. That was a goodly sum for a saddle animal in those days, but I never regretted the purchase. Kid proved to be well suited to the needs of a young forest assistant—gentle, easy to catch, and well gaited. He often carried me fifty or sixty miles on successive days as I made my rounds of the various divisions of the forest.

Kid was a remarkably sagacious animal. I remember well one November night when I was nearing Crook at the end of a long ride from Ekalaka. The afternoon had been cloudy, and with the coming of evening a low ceiling settled over the landscape, shutting out all light from the sky. The blackness seemed impenetrable. There was no road to follow, only ruts and tracks across the grasslands. And then progress was cut short by a gate, the kind referred to as a "Mormon" gate—a crude affair sometimes hard to open and close when leading a horse by the reins. Kid stood patiently while I fumbled with the wire loops, stretched the snaky wire into place, and remounted. Once away from the fence all signs of a trail were obliterated by the darkness. I knew I was in a fenced section, and that there would be another gate at the far corner. My first thought was to turn around, find the fence and follow it to the exit. But Kid refused to be turned, and I finally gave him free rein to do as he pleased. Guided by some homing instinct which man does not seem to possess, he took me diagonally across the section to the proper gate and stood while I dismounted and opened it. Soon afterwards we topped a swell in the prairie and the lights of Crook twinkled in the distance. The next day the country went to the polls and elected Woodrow Wilson to the Presidency!

With a good saddle horse and authority from Supervisor Ballinger to visit all divisions of the Sioux in order to acquaint myself with the management problems peculiar to each, life became extremely interesting. Trips were made to Ekalaka to confer with Ranger Emswiler, to Harding where Lou Shevling had his headquarters, to the Slim Buttes presided over by Ben Hubbard, and lastly to the Cave Hills, fascinating in their table and rimrock topography.

McGill, ranger in charge of the Cave Hills District, was a man slightly older than myself, whom I came to hold in high esteem after several weeks of close association. He had a Hoosier background, having been brought up in Muncie, Indiana. He and his mother had turned westward to join the tide of settlers seeking homesteads on the plains, and had filed on adjoining claims just off the northern end of the forest, quite near the Eagle's Nest Ranger Station. Mrs. McGill was a good cook and her shack was scrupulously neat. I was fortunate in being able to obtain board there during my stay in the Hills. Dave and I slept at the ranger station, but never failed to make an appearance promptly at mealtime. A motherly person, refined, and of deep religious convictions, Mrs. McGill made an impression on a young forester which has never been erased. Seen from a distance the Cave Hills appear as a flat table rising like an island from the surrounding plains. A closer look shows that this mesa has in many places almost vertical facades of buff sandstone, known locally as rimrock. In the northern section of the Hills, tortuous draws or miniature canyons lead far back into the tableland. In these twisting defiles ponderosa pines of many sizes and ages have made good growth protected from drying winds. In some places there are broad, grassy swales leading down from the mesa top, with perhaps a few green ash trees and wild plum bushes clustered about some springy spot.

Exploring this unusual bit of country on foot or on horseback was a fascinating diversion for non-working hours. Caves there were in the sand rock, and in many places high on the rim scattered beads in some rocky crevice would indicate an Indian burial spot. Looking eastward from the top of the escarpment in the late afternoon of a clear day one could watch the shadow of the mesa lengthen over the plains. How far it seemed to the distant horizon! Signs of settlement were few—here and there a homesteader's shack or soddy, a ranch house, a ploughed field, or a fenced pasture. But in spite of these evidences of civilization one felt the magic of a vast land as yet barely touched by the march of settlement.

At that time there was no detailed map of the Cave Hills, and supervisor Ballinger gave instructions to start the preparation of a topographic map of the north division which would show timber, grasslands, pockets of arable land, springs, and other features important in the proper management of the area. We were given a large planetable and telescopic alidade for the job. Initial triangulation was started from two section corners on the mesa top. The country was so open that the extension of the triangulation network was easy. We rode everywhere on horseback, Dave carrying the big planetable over his shoulder without detaching the tripod. We made use of an aneroid barometer to determine approximate elevations of rimrocks or swales. We became so interested in this project that the hours of the long summer days slipped by almost unnoticed. The completed map was accepted as being reasonably accurate, and was later incorporated in a new edition of the Sioux National Forest atlas.

The timber and grazing resources of this area have long been recognized as important factors in the economy of the region. It is also pleasant to recall the recreational assets of the Cave Hills, and the part they played in the lives of people who had few sources of entertainment, and who eagerly sought out and cherished spots of natural beauty where they might meet their neighbors at Sunday picnics and other occasions. When plums were ripe, or buffalo berries ready for picking, folks would come from all directions to gather fruit, greet old friends, and spread their lunches in the shade of some giant ponderosa pine.

Dave had recently purchased a black buggy and fancy harness for a matched pair of bays, which put him in good standing with the girls of the surrounding countryside. It was a rare occasion when he lacked company for evening rides or Sunday picnics. At the moment a pair of lovely twin sisters from a neighboring homestead seemed much in favor, and these girls were often seen in the black buggy on prairie trails, Dave, of course, holding the reins.

Wild fruit was quite an item in the diet of the plains-dweller in those days. Most sought after perhaps were the wild plums which ripened in early summer. They were usually made into jam. Service berries (Amelanchier), in spite of the large seeds and insipid flavor, were also widely used for jam. The shrub on which they grow is known in parts of Canada as the Saskatoon, or juneberry, and in the East as the shadbush. Chokecherries were most often cooked into a thick syrup—a real treat on biscuits or hotcakes. This syrup has a distinctive almond-like flavor; I know of nothing more tasty of its kind. Buffalo berries (Shepherdia) were a trial to pick on account of the thorny branches and stems, but the orange-red fruit made delicious jelly.

As a boy I had picked blueberries, wild grapes, barberries, and wildling apples from over-grown pastures in the East, and found it easy to enter wholeheartedly into the spirit of fruit-gathering picnics in this new land. I remember well the thrill of finding a plum thicket loaded with fruit in some draw leading down from. the rimrock. I remember, too, the pleasantries between neighbors as the harvest proceeded—the fried chicken, beans, home-baked bread, and cake—all shared freely with bachelor members of the expedition. Lunch was usually eaten in the shade of some box elder or green ash tree, where one could recline in comfort after the meal, exchanging stories, or just listening to the drowsy sounds of mid afternoon: the notes of sage sparrows and meadowlarks, or the gentle rustle of leaves. Many of the homesteaders who took part in these picnics had left comfortable homes in the Midwest or East to try their hand at pioneering. There might be hardships in the days ahead, but as one relaxed under the spell of a prairie afternoon he felt that for most of these people it had been a wise choice.

As summer advanced skies would be cloudless for days at a time, and the sun would beat down with blistering heat during the mid-day hours. I often started long rides late in the afternoon or in the early evening to avoid this heat, and would stay in the saddle until ten or eleven o'clock. With the going down of the sun a delicious coolness would settle over the plains; afternoon lassitude would disappear, and one felt he could ride on and on with no sense of fatigue.

*Splendid Was the Trail*

One night I saw the full moon rise just as the sun dropped out of sight in the west. At that latitude twilights are long in summer, and the hours of darkness few. It was after nine o'clock before the full effect of the moon was apparent. Then color had all but disappeared from the scene - it was a landscape of soft, luminous highlights and inky shadows. Lights at a ranch twinkled in the distance. Just how far away they actually were it was impossible to judge. Strangely enough they seemed to recede as I rode toward them.

Only the faintest glow above the northwest horizon showed where the sun had gone down; soon light would be showing in the northeast sky. If I were to catch a few winks of sleep, now was the time. But where? The need for lodging was soon met. Beside the road was a haystack protected by an enclosure of barbed wire. Through the Mormon gate the horse was led into this extemporized corral, unsaddled, and allowed to nibble for his supper. I donned my saddle slicker and burrowed into the edge of the stack. This was native wild hay, liberally shot through with needlegrass. Needlegrass is everything its name implies. In spite of the protection afforded by the slicker, the night, or what was left of it, was not passed in unalloyed comfort. At dawn sleep became impossible due to the inquisitive nature of a huge range bull which expressed his displeasure at having strangers in his domain by terrific bellows which I felt must surely be heard at the ranch I had seen in the distance the night before.

But sleeping out was not the usual practice on official trips. All of the larger ranches had bunkhouses or spare bedrooms, and forest officers were always welcome. Most often the ranch bunkhouse would be a log cabin with a dirt roof which kept the interior cool in summer and easy to heat in winter. Usually the bunks or beds had straw mattresses. Sheets and pillow slips were generally provided for guests. Almost always the heavier covers would be cotton or shoddy blankets supplemented by a number of sugans. Snuggled beneath a top covering of two or three sugans one could keep comfortably warm even on a cold night after the bunkhouse fire had gone out.

Spare bedrooms at some of the ranches were nicely furnished and evidenced the desire of the housewife to bring to this frontier some of the tidiness and comfort familiar to her in some former home. Muslin was often stretched over log walls and then covered with wall paper. Sometimes there were pieces of furniture that must have come across the plains by freight team—heirlooms too precious to be left behind when the family migrated to a new land. Beds might be adorned with priceless patchwork quilts; painted board floors brightened with braided rugs; family portraits or colored lithographs in heavy gilt frames displayed on the walls. Occasionally one would sleep on a featherbed, a most comforting experience when the country was in the grip of a Dakota blizzard.

In mid-August, I returned to the Dakota to talk over plans for nursery expansion and to look for new planting sites. It was hot. Daytime temperatures above 100 degrees were not unusual. On such days it was pleasant to shade up, under the green ash trees by the creek, or cool off in a small pond Sheriff had made by damming the stream near the station. When making all-day rides in such weather each one carried for lunch a size 2-A can of tomatoes tied in a slicker behind the saddle. The contents of the can would of course be lukewarm, but it would at least quench the thirst as we sat in the shade of a pine or juniper. There were often terrific thunderstorms in the late afternoon of these hot days.

Sheriff had designed the ranger station to suit his fancy, and I always felt it was admirably suited to the needs of that time. As mentioned before, it had one room with a screened-in porch on the south. This room would measure, as I best recall, about 35 by 15 feet. In one end was a cook stove. There was also a coal-burning heater. Ralph mined and hauled his own coal—the lignite variety—from a "mine" some miles east of the station. This coal, as it was dug, had a considerable moisture content and slacked badly when exposed to the heat of summer. It also had the usual high ash content. Some said it should be called "lugnite" for one lugged ashes all night. There was an abundance of pine and juniper available for kindling.

A rough table, bentwood chairs of Government issue, several iron cots, an Oliver typewriter and a filing cabinet for official correspondence were disposed in a rather haphazard manner in the emptiness of this large room. Water was brought from the creek in pails, which were placed on a stand in the culinary end of the room.

Ralph was a good cook and I was not wholly without experience along that line. I remember that on several occasions we feasted on delicious roast pork which came from a nearby ranch. There were always potatoes to be had, but most of the other vegetables came from cans. It goes without saying that sourdough hotcakes and biscuits were staples of diet.

The ranger station was often used for community gatherings. I shall never forget a dance that was staged during my August visit. Late in the afternoon Sheriff and I hitched up the team and drove to a ranch ten or twelve miles away where we picked up a small organ which was always available for affairs of this nature. Guests from all directions began to arrive at the station about dark. Some were on horseback, some rode in buggies or wagons. It was before the days of baby sitters: all the babies and children came with their parents.

Stout hands moved the stove out into the yard and put the other furnishings on the porch. The organ was moved in and a chair placed for a fiddler who showed up from somewhere. Children were eventually put to bed on cots on the porch or in the wagon boxes. Riding stock was unsaddled and tied about the yard or turned into the corral by the barn; teams were unhitched and given hay to munch during the long wait. The moon was near full, making it almost as light outside as it was inside the station. The crowded room would not accommodate all the dancers at one time, which made for a lot of social activity in the yard where little groups stood around and discussed the coming presidential election, the hay crop, and neighborhood matters in general. Just after midnight lunch was served by the women. Coffee was made in the yard on the stove to which a couple of lengths of stovepipe had been attached. I remember well the picture made by the sparks and billows of black smoke from fat pine surging upwards in the moonlight. All in all, it was a night one could not easily forget.

As dawn reddened the east, some of the women got breakfast, frying bacon, eggs and potatoes which, with bread and coffee, would fortify the men and boys who had to return to a day's work in the hayfields. And so they rode away, this group of friendly people, each feeling, I am sure, a little happier, and thankful for this social contact with good neighbors.

Things seemed lonesome at the Station after they were gone. Some of the men had put the stove back before leaving, but Ralph and I cooked no meals that day. We went out to the barn and slept until afternoon!

I made one more trip to the Dakota. It was in December, just before leaving for a trip to Boston on my annual vacation leave. I turned my horse out to winter pasture at the station, caught a ride back to Bowman and boarded an east-bound train there. It was January before I returned to Camp Crook.

# Tall Timber

My stay on the Sioux Forest came to an end early in 1913 when I received orders to report to the Supervisor of the Clearwater National Forest with headquarters in Orofino, Idaho, for assignment to a party which was about to make a timber survey of the Lolo Creek watershed—a heavily forested area tributary to the Clearwater River. This was an era of stock-taking, a time when it was becoming apparent that there was a pressing need for more accurate information as to how much timber was available

*Splendid Was the Trail*

to meet an accelerating demand; where merchantable stands were located, and how they could be made accessible to the lumberman. And above all, how much could be cut safely to provide a sustained yield in the years to come. Silcox, Stewart, and Mason, timber management specialists in District One, were seeking reliable data to answer these questions and others. And so were launched a series of timber reconnaissance projects which comprised an almost fabulous episode in the annals of Region One. As one who took a humble part in this program I sometimes wonder how much material garnered on those projects now reposes in the official files—data gathered by those hardy boys dubbed "timber beasts" who loved to be known for their toughness!

On the watershed of Lolo Creek was one of the finest stands of virgin white pine growing anywhere on national forest land. Intermixed with the pine was much western redcedar, Douglasfir, grandfir, and western larch. It was a stand ripe for cutting, and quite accessible for logging. I saw it first in the twilight of a day in early March; the massive trunks of pine and cedar seemed reaching for the sky above the snow-filled woods. It was with pleasant expectancy that I realized I was to live in this forest environment for the next few months.

Supervisor Fisher, known to his men as "Charlie," took me under his wing when I reported in Orofino, and suggested I ride with him to the Musselshell Ranger Station on the following day. The station was to be the base camp, and there most of the crew had already assembled preparatory to going to work.

Transportation, especially in winter, from Orofino to the station was tedious to say the least. One took the afternoon train to Greer, where lodging was had in the hotel, a venerable structure even in those days. Next morning began the long ascent of the Greer grade in a horse-drawn vehicle known as the "Stage," but better described as a cross between a surrey (no fringe on top) and a mountain wagon. From the comparatively low elevation at the river where snow seldom comes, the climb was made by a series of switchbacks to the more wintry climate of the Weippe Prairie. At Fraser, reached about dinner time, passengers were transferred to a sled. Across the prairie, through the little hamlet of Weippe, and into the darkening forest aisles beyond Brown's Creek we went until at last across a broad meadow the lights of the ranger station twinkled a welcome.

It was in the big room on the second floor of the station that I first met the men with whom I was to work: Chief of Party Fred Mason, Shaw, Richardson, Miller, Miles, Grossman, Myrick, Parker, Hastings, Fenn. Fred Mason, stocky, taciturn but friendly, was the brother of Dave Mason of the District Office in Missoula. Not given to frivolity or boisterous talk, he kept his mind pretty much on the work in hand, winning the respect and confidence of his men through fair dealing and the avoidance of playing favorites. His was a tough job and he did it well.

I had always been interested in mapping, and I was very happy to be placed in a crew which was gathering data for a topographic map. This map would include not only Lolo Creek but the watersheds of tributary streams as well: Musselshell, Cedar, Eldorado, and Mud Creeks to name a few of the most important. A surveyor, R.V. Buckner, assisted by Hastings and Fenn had run traverse control lines along the ridges separating the various drainages. These control lines were carefully surveyed with transit and stadia and platted on a sheet to give an accurate diagram of the areas to be included in the final map, and to show the approximate locations from which sketching lines would be initiated. These lines were to be run a mile apart and when completed would form a grid inside the boundaries of the control traverse. I was one of three men given the job of running these sketching lines, and I found the work interesting and challenging.

Equipment for a mapper consisted of a Forest Service box compass, a notebook with plenty of room for sketches of topography along the way, an aneroid barometer to determine elevations which were needed in the preparation of the final map which would show contours at fifty-foot intervals. Aneroid barometers are extremely sensitive to changes in atmospheric pressure due not only to differences in elevation but to weather disturbances as well. Even under normal weather conditions pressure at a given point may fluctuate in a most surprising manner throughout the day. For this reason a master barometer was kept in camp and readings noted at specified intervals—readings which would be used by the field men to make correctional adjustments in their notes.

Distances were measured by pacing—often referred to as an art. It is certainly true that to obtain proficiency in this crude method of measuring distance requires a lot of practice. One must make allowance for variances in length of step caused by climbing over windfalls, avoiding patches of devil's club, ascending or descending steep slopes, walking footlogs at stream crossings —to name a few of the obstacles met in woods travel.

The country close to the ranger station was used as a training ground for the crew—very few of the men had any previous experience in this sort of work. Until the first part of May, snowshoes or "webs" were a must for getting around, and learning to manage this footgear was an important item in the training of some of the men who were not used to winter travel in the woods. This presented no problem at all to me as I had been on many snowshoeing trips to the woods of northern New England in my school years. The tailless "bearpaw" shoe was the favorite type, although a few, including myself, preferred the conventional Canadian shoe. Some made their own webs, using pliable cedar limbs for the frames and rawhide "filling" for the mesh. As spring advanced bringing warmer weather. the snow became soft and mushy before noon, making snowshoe travel arduous and at times just about impossible late in the day. Then we went to work at daylight, so as to "catch the crust" which had formed during the night.

In about two weeks the crew had covered the country which could be reached easily from the ranger station and all hands moved to some old cabins belonging to the Musselshell Mining Company on Gold Creek. On the move each man carried his own bed roll in which were wrapped personal belongings and extra clothes. Grub was supplied to this camp, and several subsequent ones by packers who used pack boards and toboggans. Our diet was mostly ham, bacon, dried beans, evaporated potatoes, prunes, cheese, baking powder bread, coffee, with evaporated apples, cranberries and a cheap grade of chocolate for luxuries.

Conversation of an evening would invariably drift into a discussion of grub and the prospects of getting more palatable meals when pack horses could be used to supply the camps. We later found to our disgust that with the exception of fresh potatoes and a meager supply of canned tomatoes and corn there was little improvement in food when the trails were open. I doubt if a crew would stay on the job today with a cookhouse as poorly supplied as was ours at that time.

The cook was made the whipping boy. He was a rather naive individual who was out of his element among such a bunch of ruffians as we believed ourselves to be. Two of the men spent one evening whetting an enormous butcher knife which they found in an old meat house at the camp, meanwhile making remarks in a low tone of voice as to what was going to happen to a certain cook. The cook quit as soon as another man could be found to take his place. The new man's name was Wilson. He stayed with us until the end of the job.

Spring brought a few warm, sunny days, but also long periods of cloudy weather with much rain and snow. Easter Sunday found us camped at a cabin on Eldorado Creek, and a more dismal place would be hard to imagine. The cabin itself was still buried in snowdrifts. It was too small to be used for anything but a cookhouse. Sleeping tents for the crew were set up nearby, and a huge fire kept burning constantly for warmth and drying out clothes. This fire eventually melted through five feet of snow until it at last rested on the ground. Then a base of green logs was built so as to bring it up to the level of the hardpacked snow around the tents.

But at last came longer periods of sun, and the snow rapidly disappeared from the steaming woods. By mid-May much of the work could be done without snowshoes. Our first comfortable camp was at the Day Cabin. This structure, built by a mining concern, was nicely located in a dry, sunny situation, and was far more spacious than the hovel on Eldorado Creek which we had just left. Several of the men slept in the loft inside; several others, including myself, jungled out in small tents improvised from tarpaulins. A tent in front of the cabin provided dining quarters and extra room for storing supplies. A galvanized wash tub filled with water heated over an open fire gave us a chance to bathe, as well as to wash our clothes.

Join me now for a day in the field, running a sketching line by the methods in use at the time—methods now largely outmoded in the interests of accuracy and efficiency. Before leaving camp we must not forget to check our barometer with the master left in camp; it will be carried tenderly in the shirt pocket. Stopping at the cookhouse we make up a sack lunch: bread, cheese, cold bacon, a hunk of bitter chocolate, a few dried prunes or raisins. This will be carried in the rear pocket of your Filson jacket. Other pockets will provide room for notebook, pencils, compass, and tally register. When in use, the latter is best carried by a long cord around the neck.

From our starting point at the control traverse on the ridgetop it is approximately two and a half miles across a wide timbered basin to another divide where we shall tie in again to the surveyed line. We use a bearing of N 20 degrees W to establish the line we are to follow. Here the little compass in its wooden case comes into play. Taking a sight along our imaginary course we find that a big larch tree, possibly a quarter of a mile away is directly in line. Good! Now start to pace toward it, without deviation. We are using double paces; punch the tally register when you come down on the right foot. We keep track of distances in terms of the surveyor's chain— 66 feet. My pace is a trifle—a very small trifle—over five feet, or roughly thirteen to the chain.

Before we reach our tree we cross a couple of small draws. Note them on your notebook sketch map by wavy lines; ridges are indicated by symbols resembling a string of link sausages.

At our tree we find we have come almost exactly 260 paces or 20 chains. Your little map, drawn to scale, now shows two draws, two ridges, and a big tree shown by a cross. The barometer indicates we have dropped fifty feet from our starting point. Make a note of the elevation at this tree.

Now we are going down a long, gradual slope and soon find our way blocked by down timber. Here we must detour around the windfall, making proper allowances in distance to provide for this interruption in pacing. Next we come to a running stream in the bottom of a draw. No footlog in sight so we attempt to jump over it. Lose our footing and go in the water up to the knees. Nothing to worry about—it happens often. After turning the water out of our boots we sit down to eat lunch beside the stream and to jot down a few notes and do a little sketching. The bread is dry, and the cheese is mouldy on the edges, but who cares? No cup in the outfit but we can lie low enough to get a drink from the creek. It's pleasant to rest for a few minutes before proceeding with the work. Here are things which won't be recorded in the day's notes: the song of a winter wren, the insistent voice of a ruby-crowned kinglet calling from the top of a tall fir, the uncurling fiddlenecks of a clump of ferns.

Towards the end of the afternoon we start to climb a gradual slope which leads to the divide and the control traverse. On the very top of the ridge we leave a pile of stones to mark the end of our line and go in search of blazes. Yes, there is one high up on the trunk of a grandfir. On checking the station number shown on this blaze we find a slight deviation in our line has brought us a little south of where we expected to be. This is a very minor error of closure which will not affect the drafting of the final map.

Now comes the long hike back to camp, where after supper we shall work over our notes before handing them to the camp draftsman. There will be exchanges of experiences and banter during the meal, interspersed with jokes aimed at the good-natured cook.

*Splendid Was the Trail*

Timber estimating crews have been in the field today also, running lines similar to ours. An estimating line forms the axis of a strip one chain wide on which all trees are tabulated according to their diameter breast high. Tables are available to show the volume in board feet of trees of different diameters, and on the basis of this actual sampling of a small percentage of trees on a known area the total board feet of an entire watershed can be estimated with a fair degree of accuracy.

It was at the Day Cabin that Charlie Farmer appeared on the scene. Charlie was assistant to the Chief of Engineering in the District Office, and was in charge of the preparation of maps. Genial and friendly, a ready talker with a fund of stories, a prankster par excellence, inventive, a man with real ability who took unusual problems as a challenge—Charlie was one of the most talented and likable men with whom I made friends in my early days with the Service. Charlie's appearance in camp was always the signal for an open season for practical jokes. This visit was no exception.

Finding things in satisfactory shape, Charlie was left more or less free to play a few practical jokes of a minor nature. Just what they were I do not remember, but at any rate they were such as to call for retaliation, and they touched off one of the most amusing episodes of the entire job.

It was a balmy spring evening with the moon near full. We had spent an hour or so swapping yarns around the sheet iron stove in the cabin before turning in. Charlie was to sleep in one of the cabin bunks reserved for visiting guests. The cook made some preparations for breakfast and then left the cabin to go to his sleeping quarters.

When it was certain that all inside the cabin were sound asleep, two figures emerged from the shadows into the moonlight. One carried a flat rock, the other some fir boughs. The latter went inside the cabin and stuffed the boughs in the stove. When it was certain a good rousing fire was kindled, the other man climbed on the cabin roof and put the rock over the stove pipe. Then the cabin door was wired shut from the outside.

It was only a matter of minutes before there were signs of life inside the cabin. Smoke was pouring out from beneath each shake, which made a weird sight in the moonlight. Coughing and sputtering mixed with intemperate language and pounding on the closed door created a pandemonium. The tumult attracted all who were sleeping outside the cabin. The climax occurred when someone inside had the presence of mind to take out one of the small windows through which exit, amid clouds of smoke, the men made their escape, one by one.

While camped at the Day Cabin, two of the men—I think they were Shaw and Miller—killed a bear, which provided fresh meat to vary the ham-bacon diet. We had it fried, boiled, in stew, and ground to hamburger. We had it for breakfast, lunch, and dinner. Never did any of us care if we ever ate bear meat again!

Those who know of the strong resemblance between the bones of the front leg and paw of the bear and the skeleton of the human arm and hand will appreciate the effect that was achieved by nailing this skeletal appendage of bruin over the cabin door as a hint of what would occur to a certain ranger should he appear in camp! This ranger was supposed to be the man responsible for the poor quality of the grub furnished us. We later learned the supervisor's office and not the ranger was to blame. But anyway, word of our doings reached the ranger via the grapevine, and when he was forced to visit camp he brought the supervisor as body guard. Over the same grapevine came the report of his remarks on getting safely back to the ranger station: "They're a tough lot, those boys, a bunch of young cutthroats. They say they'd kill a man if they caught him alone, and by heck I think they would!" Poor old Bob Snyder, kindly and lovable, a man who tried to do his job as he saw it. May we be forgiven for the rough time we gave him in those far off years!

There is much more of an amusing nature that could be told, incidents that will come to mind as I re-read this brief narrative: the evening song rests, the terrifying groans of the boss as he struggled with a nightmare, the no-see-ums and the mosquitoes that came in due season, the evening expeditions after venison, the side camp excursions to remote areas, walking the footlogs over swift water, bucking the wet brush after a rain. There were few, if any, dull moments that I can recall.

Early summer found us working in the Mud Creek drainage, the most southerly part of the Musselshell area. Here the topography was not as diversified as farther north. There was much comparatively level country and few landmarks. Mapping became more of a problem due to the inefficiency of aneroid barometers in measuring very slight changes of elevation. Thunderstorms which developed almost every afternoon made it still more difficult to determine elevation with even a fair degree of accuracy. We often talked among ourselves of better methods of mapping and, sure enough, a better method was being devised, although we did not know about it at the time. The next year a little hand level known as an Abney, with which the degrees of slope could be ascertained, and the steel tape for measuring distances were put into use and found to give more accurate results than the old system.

Probably no man on the job played a more important part than Jim Yule, the camp draftsman. A man of infinite patience and great resourcefulness, to whom regular hours of work meant nothing, Jim made a notable contribution to the success of the Lolo Creek job.

The work came to an end the last of September. Before the windup, I had been assigned to a reconnaissance crew on Cedar and Cougar Creeks of the Coeur d'Alene Forest as camp draftsman. With the completion of this project late in the fall, my timber cruising experience ended and I began a period in the Missoula Office of Engineering as a topographic draftsman, an assignment which brought me great satisfaction.

# The Lost Battalion

The tragic holocausts of 1910 had brought a greater realization of the ever present threat to our forests from fire, and the magnitude of the task which lay ahead in controlling this enemy. Field-going men were on constant call during the summer months to carry on the campaign; members of the office force were often sent out to help in the capacity of timekeepers, commissary men, or camp bosses.

In those days there were no smokejumpers who could be dropped on a fire in remote back country in a matter of minutes after it was reported. Great dependence had to be placed on the stamina of the rugged boys known as "smokechasers" who take off on foot to a fire, laden with tools, a flashlight for night travel, and rations for wilderness survival. The speed with which a smokechaser reaches his objective can mean the difference between a small controlled spot fire and a big blaze which may spread to mammoth proportions. These big fires may burn for weeks and be finally controlled only when the fall rains come. Battling a fire of such size always has been and always will be a costly undertaking.

Firefighters working for the Forest Service have usually been well cared for. In the years when there were fewer roads into the back country, and air delivery was yet to be developed, getting men and supplies to remote camps posed a real transportation problem. This was the heyday of the Forest Service mule and the hardy breed of men who handled them—the packers.

The summer of 1915 was not considered a bad fire season, but there were, nevertheless, a number of stubborn blazes which kept the personnel of the various forests on their toes. The Clearwater Forest, with headquarters in Orofino, Idaho, had sent men and supplies to one of these fires, which was burning on Isabella Creek, a tributary of the North Fork of the Clearwater River, in the magnificent white pine forest which clothes so much of that part of northern Idaho. It was a remote locality in those days when the North Fork was roadless; the nearest road-end was at Headquarters, an outpost of the Northern Idaho Forest Protective Association, about twelve miles from the historic hamlet of Pierce. From Headquarters to the fire was a long hike; over Bertha Hill to the North Fork and then to the mouth of Isabella Creek, up which the way led probably three miles to the fire. Crews going to the fire were accompanied by a packstring and camped out one night along the way.

Reports trickling in to the office of fire control in Missoula were scant; repeated calls to Supervisor Willey in Orofino brought only the terse reply, "We are making progress." Then light rains came to dampen things up a bit, but still no favorable word from Isabella Creek. The vague words over the telephone, "We are making good progress," brought little satisfaction to those in the District Office who were watching mounting costs in connection with this illusive conflagration. The crew working on the fire was dubbed, "The Lost Battalion."

Today we wonder how we ever made it without the airplane and radio in forest fire control. But two essentials we never lacked—dedication and persistence. The old timer would use the homely phrase, "Just plain gutsy!"

It was about six o'clock of an evening in the latter part of August that the following message reached me from the Office of Fire Control: "You will report for duty on the Isabella Creek fire on the Clearwater Forest. Proceed to Superior on the next westbound Milwaukee train, which leaves Missoula about eleven o'clock. C.B. Swim will be in charge of your party and will be at the depot to arrange for your transportation. He plans to reach the fire by the shortest route across country from Superior. Take a bedroll with you. Plan to be away for an indefinite period."

Frankly, I was pleased to receive these orders. Here was adventure in the making. Of late I had spent much time attending to the drab details of ordering supplies, rounding up firefighters from the saloons and jungles and making sure they were placed on trains for various destinations. All this I could now leave in other hands for a spell—maybe for weeks—and again fill my lungs with mountain air.

Swim was on hand at the station and the train was on time. A young man, quite obviously of Scandinavian background, was the third member of the party. Although he had no previous experience in fighting fire, he proved to be a good man on the job. I admired the way he measured up to his responsibilities from the start of the trip, which turned out to be a rough one. I doubt if he had ever ridden a horse before.

We reached Superior in time to catch a few hours sleep in the Charette Hotel, a well-kept and homey sort of place. Swim was out before six o'clock and informed us at breakfast that he had hired four horses, three to ride and one to pack. He was the self-appointed packer, and soon had our beds and other gear slung on an old-fashioned sawbuck packsaddle and cinched down with the conventional diamond hitch. Sack lunches were carried rolled in slickers tied to our saddles. "We'll be in camp in time for a good evening meal," said Swim.

Our objective the first day was Chamberlain Meadows, where Fred Wilfong, employed that season as a packer, had his headquarters in a comfortable, though primitive cabin. It was a foregone conclusion that Fred would gladly furnish us with a hot meal and a place to spread our beds. There, too, we would find a corral and feed for the horses.

I think the three of us were sleepy that morning; the ride up Cedar Creek from Superior and Iron Mountain seemed long and tedious. "Never hurry a packhorse," was one of Swim's mottoes, oft-repeated. He practiced what he preached and we jogged along at a sub-normal pace, interrupted frequently by stops to readjust the pack. Lunch was eaten hurriedly as we didn't want to leave the horses standing to be pestered by flies. We had a lively time with yellowjackets, which were attracted by our sandwiches.

Late in the afternoon we began to wonder how soon we would reach our destination. We had been climbing steadily for some time. We seemed to be headed for Illinois Peak, which loomed large on our left. I called this to Swim's attention, but he remained firm in his conviction that we were still on the trail which would take us to Chamberlain Meadows. But a half hour later it was all too apparent that we had missed the way; our destination was surely far to the west of our present position. We were tired and hungry, and the sun had already dropped behind the ridges and soon would be gone entirely. After a consultation we remembered that some distance back we had noticed an unsigned trail leading off to the right and we now felt certain that was where we had missed the way, by keeping straight ahead instead of turning right. On this assumption we retraced our steps and found our surmise was correct. It seemed a wearisome journey to the Meadows, which we reached about eleven o'clock; the little cabin was barely discernible in the moonless night. At least one of our horses recognized familiar ground and let out a neigh which was answered by one of Fred's animals in the corral.

Wilfong, who had retired at a decent hour, rubbed the sleep out of his eyes, welcomed us with a smile, fired up the cookstove preparatory to making hot biscuits, showed us the horsefeed, helped us unpack, and gave us a place in the cabin where we could spread our beds. In the years which followed I came to know Fred as a man who was always smiling, and I have never forgotten the time he played host to us so cheerfully at that little wilderness outpost in the middle of an August night. The biscuits he baked for three hungry men were the best I ever tasted, so I thought.

Pole Mountain Lookout was our objective next day. It was an interesting ride, the trail in places traversing stands of lodgepole pine which had re-clothed country burned many years before. Many of the older trees on the ridges, fantastic and grotesque in their twisted growth, would have made fine subjects for the camera, had we taken time to stop. We made good progress that day and reached our destination early enough to enjoy a hearty evening meal cooked by the man on duty at the lookout.

Next day we reached Mallard Lake early in the afternoon, to be greeted by Ranger John Durant, and the man in charge of Mallard Lake Lookout, Fleming Stewart, a man whom I came to know well in later years, and to admire as one who always gave the very best effort to any job assigned to him. Here on the lookout he had learned to operate an heliograph, his chief means of communication with the outside world. "Fine," said he, "when the sun is shining, but no good in cloudy weather." John Durant I had known previously when I was with the timbercruising party in the Musselshell country. He and his brother Exie operated a creamery at Weippe, Idaho, but I always had the feeling that Johnny much preferred the life of a woodsman—and a fine one he was—to the routine of juggling cream cans. Most woodsmen that I have met and worked with have been kindly men that one could admire for their feats of skill in the timber, and Johnny Durant was no exception.

John said the latest word from the fire indicated it would soon be under control and suggested that Swim take over supervision of the mop-up. Johnny seemed a little concerned about work which had been piling up in his district during his absence on the fire. Accordingly, it was arranged that Swim and the timekeeper go down to the fire next morning and that I go with John to a camp on the North Fork where the crew of men originally hired in Orofino as firefighters was building a much needed trail. This trail project was in the charge of George McKinnon, an able and well-liked woodsman and trapper who had built himself a home on the North Fork not far below the mouth of Isabella Creek. To use fire funds for trail build-

*Splendid Was the Trail*

ing or other needed improvements was not an uncommon practice in those pioneer days of the Forest Service—a practice which kept men employed on constructive projects pending the time a fire could be pronounced thoroughly out and safe to leave.

Here was an outstanding example of such practice. Said Johnny, "How do we know the fire won't blow up? We can't let these men go until she's dead out. Maybe when the first snow comes. Think of the trail we can build in the meantime! I hope the District Office stays out of this," looking at me.

Yes, I was a representative of the District Office, but in this primitive region, roadless and sadly in need of trails—a country which seemed far removed from the routine of desk work, reports, and red tape—I found it not difficult to appreciate the viewpoint of men like John and George who spent most of the year fighting the brush on foot or on snowshoes, and to condone their disregard for the niceties of Forest Service fiscal procedure. None of us who were regular employees of the Forest Service in that locality were naive enough to believe that Supervisor Willey and his staff were not aware that a much-needed trail project was in progress in a grand way on the North Fork. It proved to be an enthusiastic crew; where the men could see constructive progress there was a will to work.

Johnny was in prime condition for the hike, a long one, from Mallard Lake to the river. Clad in a wool undershirt, over which he wore suspenders which held up stagged pants, and shod with custom-made loggers, he looked the part of a northern Idaho woodsman rather than a Forest Service ranger. I doubt if he ever owned a uniform or the regulation Stetson hat. But the keen eyes peering out from beneath the brim of a floppy hat, the rhythmic gait which carried his wiry frame uphill and downhill, over windfalls and through thickets of ceanothus brush, the quiet way in which he gave decisive orders to his men, marked him as a man one could depend on to cope successfully with back-country problems. "It'll be a tough hike to the North Fork," he said, "but I think you can make it." Right then and there I made up my mind that come what might, I would make it.

Water was scarce on the ridges we followed, and the day was hot. "Good day for a blow-up," said John. There was no trail; we kept to the high country picking our way to take advantage of open slopes and ridge-tops where there was not much growth but beargrass and scattered clumps of ceanothus. Occasionally we caught glimpses of smoke from the fire below and remarked it didn't look too serious, unless a hot, drying wind should come up in the afternoon to cause it to crown and spread. We carried meager lunches of sandwiches and bitter chocolate which we ate about midday by a trickle of water which we found in a draw. Finally we reached the last descent, a timbered hillside down which we scrambled to the North Fork.

It was getting dark by this time and I was so tired I felt I could fall asleep propped against a tree. John's white undershirt caught a vestige of light under the dark crowns and made a welcome beacon to follow. Soon we came to a newly constructed piece of trail which eventually took us to camp, a welcome sight with light from gas lanterns pushing back the darkness. Too tired to eat, I crawled into a firefighter's bed and listened to the subdued voice of the North Fork, which seemed to murmur, "Well, you *did* make it!"

Next morning I met McKinnon for the first time. His cabin was one of the finest log structures I have ever seen. The interior was finished with cedar panels split out with a frow and smoothed with a hand plane; the kitchen cabinets were of superb workmanship which would have been a credit to any carpenter; on the puncheon floor were trophies of many hunting expeditions. George gave me, before I left, the hide of a mountain goat, a memento of which I was very proud.

McKinnon was a delightful man to be with, and as my duties as timekeeper were light, I had the opportunity to listen to his tales of hunting and fishing; his accounts of encounters with bears; his many other experiences as a backwoodsman and prospector. He was a clean liver, and always stayed sober on his infrequent trips to town. His conversation was singularly free from coarseness and profanity, and his kindliness and gentle manners made him acceptable in any company. I last saw him several years later when he was working as foreman for Potlatch Forests; he had married and posed for a picture holding the hand of a tiny daughter, a shy little girl of whom he was very proud.

When I took up my duties as timekeeper, there were about fifty men in the trail crew. When the fire on Isabella Creek was definitely out a week or so later, most of the men came down to join the trail-builders. Supervisor Willey and other members of his staff came out from Orofino to spend a day and see how the work was progressing and were delighted with what was being accomplished. Nothing was said about terminating the job, and I was reasonably certain that George was given a long wink which would amount to orders to keep things moving until further notice. At any rate we all reflected the happiness of the supervisor at the amount of trail being built with fire funds. It was a woodland Shangri-La; we had good cooks and plenty of grub, the hot weather had eased up a bit, and we could all see tangible results from our labors—a big factor in boosting the morale of a crew of men.

Then disconcerting news was relayed to us one evening; a member of the District Office was on his way from Missoula to see what was going on, and would reach our camp the following day! Apparently the powers-that-be had not been kept fully informed as to the status of the fire on Isabella and were appalled at the mounting expense in connection with a blaze that should have been controlled days ago. George and I talked over the matter and decided there wasn't much to worry about as far as we were concerned; the responsibility lay with the Orofino boys, not with us.

Weary and saddle-sore from an unaccustomed ride on forest trails, the man in question arrived in camp the next evening in time for supper. This "inspector," as we jokingly called him, proved to be Edgar F. White, in charge at that time of District One planting activities. He was an extremely affable and tactful man, well-liked because of his friendly disposition and the good sense he displayed in dealing with administrative problems. He immediately sized up the situation, admitted the need of more good trails in the locality, commended our zeal in attacking the problem, and refrained from criticizing too strongly the irregular way by which we were financing the job.

White had ridden many hobbies; at the time of this visit to the North Fork he was making a study of medicinal herbs which might be growing in Northern Idaho. In such free time as I had I joined him on exploring trips in the vicinity of camp, where he found several species of commercial importance. It was pleasant to roam in the shade of the white pine forest and talk of the possibilities of picking up a little extra cash by selling plants to Chinese herb doctors or exporters to the Far East. A few years later White and I established a backyard bee business when honey was at a premium during World War 1. This venture I took over when my partner was transferred to the Forest Products Laboratory at Madison, Wisconsin.

These halcyon days on the North Fork finally came to an end when we were told to break camp and to proceed to Pierce where a final audit and OK of the timeslips would be made and the men shipped to Orofino.

"I think we can make it to Pierce in a day," George said. "It's about thirty miles to Headquarters where we can catch a ride to Pierce, but we can shorten the distance quite a bit if we take a boat down to where the Bertha Hill trail leaves the river. It will be a two-day hike for the men; they will spend a night at the lookout. You'll have plenty of time to get organized in Pierce before they get in sometime in the middle of the second day."

Early in the morning we shoved our boat away from the bank into the swiftly moving North Fork. The water was low at the close of the dry summer and there were many jutting rocks to be avoided, but George was a good riverman and brought the craft to a safe landing six or seven miles downstream. We pulled the boat out of the water as far as we could, made it fast, and started up the trail which would eventually take us to Bertha Hill and beyond.

It was a beautiful September day, just brisk enough to make the easy climb quite enjoyable. We weren't slowed down by superfluous baggage—my packsack held only a shaving kit, the goat hide, and a few essentials of clothing. George took a slow, steady pace, stopping very rarely for a breather. I took note of his swinging gait, common to most woodsmen. Many phenomenal walkers whom I have known—Robert Marshall, for instance—used this rhythmic swinging gait which seems to carry the hips forward alternately with a relaxed rolling motion. Some call it "the woodsman's walk." Whatever we call it certainly covers the ground.

At the lookout on Bertha Hill we ate lunch and then started down the trail to Headquarters, an outpost of the Northern Idaho Forest Protective Association, where we felt sure we could get transportation to Pierce. The day was fine but we noticed cirrus clouds coming into the sky which drew the remark from George, "The fall rains could start anytime now." How true those words proved to be!

It was after dark when we reached Headquarters, consisting at that time of a few buildings grouped in a clearing in the dense white pine forest. The warden gave us a cordial greeting and introduced us to the cook who set out a meal for us. Later we made arrangements for transportation to Pierce in a Model T Ford, a distance of some twelve or fifteen miles; long miles they seemed on the tortuous dirt road, the light car careening from side to side as the driver dodged the chuck holes, half-buried roots of trees, and deep ruts which suddenly emerged out of the darkness into the dim light cast by headlamps which fluctuated to suit the speed of the engine. At last a few lights and then the little town of Pierce—a ramshackle place with dilapidated buildings strung along the unkempt main street. Once the capital of Idaho, and famous for the placer diggings thereabouts, it had lost all appearance of a metropolis and could qualify for a leading place in any parade of ghost towns—so I thought as I saw it by daylight next morning, as rain fell in a steady downpour which turned the unpaved main thoroughfare into a sea of mud.

At that time the Forest Service had leased a house which had formerly been an eating place belonging to a Chinaman, and here a fireguard had his headquarters during the summer. Certain repairs had been made to the old structure to insure its being weatherproof. A front room equipped with a desk and telephone, an Oliver typewriter, and a USDA calendar lent a certain air of importance to the establishment, and served as a place to receive callers and to transact official business. In the rear, in what had been the kitchen, was a room which would sleep several men in an emergency. Here I unrolled a firefighter's bed, threw a few sticks of wood into the stove to lessen the dampness, and fell asleep to the drumming of the rain on the galvanized iron roof.

My job at Pierce was to check and countersign the timeslips which showed the pay each man had coming and to make deductions to cover the cost of items they had purchased at the camp commissary. The men began to arrive in the middle of the forenoon and a sorry sight they were. Many of them had spent the night without much shelter at the Bertha Hill lookout and were thoroughly soaked before reaching Headquarters, where they were given a ride to town. There was considerable grumbling but this quieted down after a hot meal at the restaurant up the street. Eventually transportation showed up to take the crew to Orofino where they were paid off by a deputy paymaster.

A day later I left Pierce for the return to Missoula by stage and train. At Pardee, a whistlestop between Kamiah and Greer on the railroad which runs up the valley as far as Stites, I spent a night with old friends who had a wonderful melon patch. The cantaloupes were ripe, and I picked all I could carry in my packsack. Having time between trains in Lewiston, I took a long walk through the Clarkston orchards across the Snake River in Washington and bought a basket of quinces, a fruit I hadn't seen since I left the East. These were quite a novelty when I passed them around among friends in Missoula. So ended the part I played in the episode of the Lost Battalion!

*Splendid Was the Trail*

Wylie's Peak

# Year of the Armistice

The summer of 1918 found the personnel of District 1 with badly-frayed nerves. World War One was in full swing, and many of our key men were in military service, leaving a much-reduced force to carry on necessary work, both in Missoula and on the forests. Fires, some of them believed incendiary, were burning in many places. Recruiting firefighters posed a real problem. Members of the far-left organization calling itself the International Workers of the World, popularly known as "Wobblies, "were fomenting all sorts of labor trouble; they had the Forest Service over a barrel—or so they thought. Food was rationed and supplies costly, which made it difficult to feed our crews in the manner to which they had become accustomed. It was under these conditions that I was asked to take fifty men over the trail to a fire that was burning near the confluence of Bear Creek with the Selway River the first part of August. At the Bear Creek Station my responsibility would cease when the men would be turned over to John (Jack) Clack, assistant forest supervisor of the old Selway National Forest, with headquarters at Kooskia, Idaho.

The agitators were easily spotted among the motley crew which climbed aboard the afternoon train which in those days ran every week-day from Missoula up the Bitterroot Valley. Loud-mouthed and demand-ing, it was evident they were out to make trouble—it was suspected that they might even be professional saboteurs. Be that as it may, the more docile members of the crew seemed wholly subservient to their whims, and willingly took lessons from them in grumbling and trouble-making.

Assistant District Forester Glen Smith met the train at Como, a whis-tlestop a few miles below Darby; there trucks were waiting to take men and supplies to the lower end of Como Lake, where transfer would be made to a power barge for a four-mile trip to a packers' camp at the beginning of the Rock Creek trail. I think Glen was a little startled at the appearance of my crew; with kindly assurance he shook my hand and wished me Godspeed.

After loading equipment and supplies on the barge, which was moored at an improvised dock near the earthen dam which impounded the waters of the lake for irrigation, it became apparent that there would not be room for all hands on the boat. A few sat on bales of hay which were piled on the deck, but the greater number took to the trail and reached camp before the slow-moving craft had touched shore on the other side. The craft was met by several packers who were waiting to manty up equipment and supplies in readiness for an early start the next day. A cook and flunky, who with the packers were local boys, had a fire going preparatory to getting supper. Pack animals, mostly horses which had been picked up where available, were tied to trees on the edge of camp. These animals knew well what lay ahead, for they had shuttled several times that season between the Bitterroot Valley and the Selway Riv-er over the Rock Creek trail which we were to follow.

That we could expect little or no cooperation from this bunch of men in the days ahead was clearly indicated by an incident that occurred dur-ing the evening meal. Sugar at that time was rationed rather drastically and was not always easy to procure. A hundred pound sack had been allotted to this expedition and was placed under the watchful eyes of the cook. An amount considered ample to supply the needs of those who wished to sweeten their coffee at supper was placed in a dishup pan where all could reach it. Soon the pan was empty and profane cries filled the air calling for more, and threatening violent reprisals on the cook if it were not forthcom-ing. The packers and I finally quieted the disturbance; there were sullen looks and low growls, but the sugar sack was safely tucked in for the night.

Right then and there the packers and I became good friends.

As we left camp the next morning the men were cautioned not to get too far ahead of the pack string—in fact, to stay behind the horses. This advice met with little response; each man carried a lunch and felt quite independent and able to travel at his own pace. A few were soon far ahead, and out of touch with the main party, while others who were barely sober after lengthy sprees in the bars of Butte and Missoula found it difficult to keep up. The day was warm, and the trail dusty. Whenever a stop was made to adjust the packs, flies of several species tormented the horses. I kept some distance in the rear to avoid the dust, but never so far behind that I could not hear the sweet note of a Swiss bell attached to one of the mares.

Late in the afternoon camp was made on an open hillside where the vegetation was mostly ceonothus and beargrass, with an occasional clump of lodgepole pine. Nearby was a small stream whose flow was sufficient for our needs. Stragglers began to show up from the rear, but it was soon apparent that a goodly number of the men were far in advance. About sundown a group of three returning from this vanguard said the main party was a long ways ahead—possibly four miles—and wanted to know when they were going to eat. The colorful reply from the packers indicated that no special service would be forthcoming for those who had out-walked the packtrain in disobedience of orders. A few more stragglers appeared in the next hour and regaled themselves with sandwiches and coffee, provided by the kind-hearted cook. Possibly ten were missing at final roll call.

Shortly after daylight the camp was thrown into an uproar by the return of these missing men, vowing vengeance on the dirty little timekeeper who caused them all this misery and demanding he be turned over to them to be flayed alive. I shall always remember the magnificent way in which the packers, my staunch friends, handled the matter. One had slept with his gun beside him. "Just in case of bears," said he, and he held it in the crook of his arm as he faced the disturbers. The other two packers and the cook armed themselves with such weapons as came handy, and left no doubt that there would be cracked heads and worse in case of insurrection. Things finally quieted down, and the growling men were considerably mollified by a warm breakfast. Later several conceded rather shamefacedly that after all I was not really responsible for their going supperless and sleeping without covers in a vast howling wilderness. I think some of the men really had a dread of wild animals on the prowl.

It was on the afternoon of this day that I met for the first time two men with whom I was to have close acquaintance in the years to come—Jack Clack and Bill Bell.

Jack, in his role of assistant supervisor of the Selway, played a very large part in the fire control activities of that forest, and had come to Bear Creek to help Bell at this critical time. The fire being just about dead, he decided to take the men I had brought in down river to a trail job that was underway and hold them there as an emergency crew until the danger from lightning storms was definitely over.

Hatless and with sleeves rolled up when I first saw him, Jack was supervising the construction of a "Dago" oven such as he had used many years before when was building railroads on the Canadian prairies. It was a dome-shaped affair built of rocks and earth, which could be heated to the desired temperature for baking bread by building a fire inside. Clean-shaven and tanned, resourceful and energetic, and with always a twinkle in his eye, Jack Clack played well the role of a great out-of-doors man.

Sizing up my men in the light of what the packers and I had told him, he decided a few more miles on the trail that afternoon would serve to keep their minds off their troubles, real or imaginary, and after a rest period gave the word to move. It was with a great feeling of relief that I watched a curtain of dust obscure the retreating forms as the procession followed Clack down the trail.

Bill Bell had the same feeling when we saw the men disappear. Together we went into the cabin which served as the Bear Creek station. It had a puncheon floor skillfully hewed with a broad axe, and a roof of cedar shakes. Several summer employees used the cabin for sleeping quarters, and now that the fire crew cook had gone down river, meals would be prepared and eaten at one end of the big room. Bill was a good cook indoors or out and took over the job of preparing meals. In fact Bill had all the skills required of a woodsman plus the ability to get along with horses. Years later he was one of the top foremen at the Forest Service Remount Depot, a position he held until his retirement. He loved mules and horses—and I feel sure his charges sensed it. "Kindness goes long ways with a mule," Bill often told me.

"Never go out at night without a flashlight," cautioned Bill. "This place is a hangout for rattlesnakes."

And so I learned shortly. That very evening in crossing the yard to facilities in the rear I encountered a good-sized rattler beside the walk. He had coiled and was sounding his note of alarm when I spotted him with the flashlight. Hurrying back to the cabin, I told Bill who grabbed his 22 rifle and went out to look for the snake, who had made his getaway by the time we returned to the spot where I had seen him.

"Lots of them in this country," reiterated Bill. "Sometimes you can hear them under the cabin floor."

And indeed, lying in bed on the floor and kicking our heels on the planks we did hear them buzz at times. A dozen or more were killed in the yard that summer.

Bill liked nothing better than to ride the trails of his district on horseback, and a day or so after my arrival at Bear Creek he decided to make a necessary inspection trip, and invited me to go with him. A heavy thunder shower with copious rain had made the woods safe from fire for a few days, so leaving three or four men to do some work in the vicinity of the station and to make daily reports over the telephone to headquarters in Kooskia, we started for Gardner Peak to investigate its value as a lookout point.

It was a bright, clear morning, with the moist woods fragrant after the

*Splendid Was the Trail*

heavy rain. Bill rode ahead, leading a packhorse which carried our bedrolls, a tent, and food for several days. There wasn't much of a trail to follow, but the forest was more or less open, with surprisingly little windfall. The top of the hill proved to be anything but a peak, and to get a view of the surrounding country we had to climb Douglasfirs growing on the ridgetop. As there was no water this high up, we camped at a spring some distance down the mountainside. There each horse was given a feed of oats, hobbled, and turned loose to graze on what feed might be available in the vicinity.

Bill was adept at making camp, and I admired the skill with which he arranged things with an eye to expediency and comfort. He dug a short trench about a foot deep, over which he laid two small green logs spaced so as to support a kettle, a frying pan, and a coffee pot. A fire kindled in the trench soon had water boiling and coals which could be raked out to bake bannocks or frying pan bread. Bill had prepared a small sack of biscuit mix before leaving the station by adding the proper amounts of salt and baking powder to the dry flour; lard for shortening was added later, and then water to make a dough of proper stiffness. The lard was usually worked into the dry flour with the fingers before the water was added. The biscuits, almost as large as the pan, were first cooked on the bottom, and then browned on top by tipping the pan to an upright position in front of the fire. Eaten with plenty of butter, or possibly larded with bacon grease, these bannocks would satisfy the hungriest man; spread with strawberry or fig jam they made a satisfactory dessert. There was always plenty of ham and bacon on pack trips; canned tomatoes were a staple; coffee was made stout and taken black—a beverage I never learned to enjoy. I was often ribbed for indulging in canned milk thinned down with water.

To waken early in a mountain meadow on a bright summer morning is an exhilarating experience. One may listen for a few moments to a bird call or the distant tinkle of a horse bell before rolling out, but there is no desire to linger in bed. Minutes of a day like this are too precious to waste; with each breath of smog-free air one is more certain than ever that he wouldn't trade places with any city dweller of his acquaintance. It was on such a morning that we started the day on Gardner Peak.

Game had been around the waterhole in the night: I noticed deer tracks in the soft ground, and the print of a larger hoof which might have belonged to a moose. On both sides of the little rill were patches of yellow monkey flower and the blue stickseed which resembles forget-me-not. There were clumps of white valerian, sometimes called false heliotrope, deep lavender blossoms of wild onion, and farther from the water bushes of shrubby cinquefoil and pink spirea. It was a pleasant spot, moist and cool before the heat of the sun had dried the heavy dew which glistened everywhere on the vegetation. We felt fortunate to be there to enjoy it all—definitely the first human visitors to the spot that season, and possibly the last for some time to come. Love of the wilderness comes naturally to most foresters.

The builders of Wylie Peak Lookout had an eye for the spectacular.

*"Rimrock Sentinel," Swan titled this picture. He considered the scenes it had witnessed of Native Americans and early whites, and pondered the power of winter storms that shaped it physically.*

All photographs in this section were taken by Kenneth D. Swan, and appeared in the original edition of this book. With one exception, they are from the collections of University of Montana Library, and are reprinted here with permission. The library's catalogue number appears by each image.

91-78

93-3579

*Above: Western white pines in northern Idaho.*

*Top: The main street of Camp Crook, South Dakota, in August 1912.*

II

91-336

93-3574

A 1926 view of what is now Medicine Rocks State Park, 30 miles south of Baker, Montana.

*Cutting ponderosa pine north of the Kootenai river near Libby, Montana, 1921.*

*"The Goblin's Armchair" in what became Makoshika State Park near Glendive, Montana*

IV

*Above: The crest of the Mission Mountains in Montana.*

*Top: Jack Clack fries up bacon for breakfast in the wilderness.*

93-3569

*An early-day smoke-jumper aims for Sleeman Creek on the Lolo National Forest in Montana.*

93-3577

*As a smokejumper heads for the same fire, his small "pilot chute" pulls open his main chute.*

*The Half Moon forest fire of August 1929, illuminated by the setting sun and viewed from Coram Ranger Station, Flathead National Forest, Montana.*

*Bob Johnson piloted this Travelair (below) to Big Prairie on the Flathead National Forest, Montana, loaded with back-country supplies.*

*Above: Dick Johnson, with his brother Bob, ran Johnson Flying Service of Missoula, Montana and was an important part of creating the first smoke-jumping program.*

*Facing page, top: Whitebark pine on the Beartooth Plateau, Custer National Forest, Montana.*

*Bottom: Big Salmon Lake in the Bob Marshall Wilderness Area of the Flathead National Forest.*

93-3571

*Above: Bob Marshall examines a ponderosa pine girdled by Indians but still flourishing, near Florence, Montana in the summer of 1928.*

*Facing page, top: Riders in the Bob Marshall Wilderness Area approach their night's camp in Pearl Basin.*

*Bottom: Leaving camp at Wounded Man Lake on the Custer National Forest.*

*A tarn well-hidden in the Spanish Peaks Wilderness Area of the Gallatin National Forest, Montana.*

*At timberline in the Mission Mountains Wilderness Area, whitebark pine acts as a natural weathervane.*

*A fire lookout in the summer of 1943, Mrs. Edith Averill stayed with her granddaughter at Cayuse Point on the Lolo National Forest, Montana.*

*Above: Snowshoes are the vehicle of choice for a sunny winter day's hike on Mount Stuart, Lolo National Forest.*

*Facing page, top: In the Hilgard Country of Gallatin National Forest.*

*Bottom: Storm Pass in the Anaconda-Pintler Wilderness Area of the Deerlodge National Forest, Montana.*

*Above: Jack Clack in the "Hall of the Old-Fashioned Ladies" on Mount Stuart, Lolo National Forest.*

*Top: A moose enjoys early morn in Hoodoo Lake, Clearwater National Forest, Idaho.*

Perched on a rocky knob high above a meadow strewn with boulders which had tumbled down from the cliffs above, it was an attention-getter. The cabin provided a wide view in all directions, and was considered at that time one of the most strategic observation points in that locality. There the man on duty greeted us with the glad tidings that an armistice had been signed, and that the war was undoubtedly over. The word had just been relayed over the telephone, presumably from Kooskia. Later we were to learn that this report was erroneous—the final armistice was not consummated until November 18. But we had our moments of elation before we finally learned that a mistake in reporting had been made. Who started the rumor we never found out; we hope the error was unintentional, for it seemed a cruel hoax at the time. Today we think back to the days when there were no radio sets which could be tuned in for the latest minute-by-minute news!

Mapping the forests has always been an important activity of the Forest Service. It seems a far cry from present-day refined methods to those in use the early years when maps of large areas were sometimes little more than rough sketches with some parts virtually blank or filled in with unreliable data—mostly guesswork. Discrepancies had been noted in the map of the Running Creek drainage, and corrections that should be made were very apparent. Feeling certain that I would have some spare time while in the Bear Creek country, the Office of Engineering had asked me to take readings with a compass from high points in the Running Creek area, and by triangulation to ascertain their approximate location. Bill was eager to give me assistance on this project, and of course I was more than happy to have his help. So after our premature Armistice Day celebration, we started south into a country as nondescript and trailless as any I had ever traversed with horses.

Several days were spent in climbing to high points and taking readings with a standard Forest Service compass. The errors in the map became more and more glaring as this work proceeded. The job was not easy but we thoroughly enjoyed it; we met the challenge of Idaho brush, climbed over windfalls, enjoyed the coolness of pockets of timber, lingered to enjoy the views from summits we scaled. Today aerial photography has taken some of the hard work out of mapping, but it has also eliminated some of the joy of achievement under odds.

For several days there were thunderstorms in the afternoon, causing Bill some anxiety about possible fires. We both agreed that it would be best to return to Bear Creek as soon as possible.

One shot from a high point in the area was needed to complete our triangulation, and accordingly we decided to make a sidetrip from camp to a certain summit early in the day and then travel home to Bear Creek in the late afternoon. And so we had an early breakfast on this last morning of our trip, got our gear in shape for easy packing, tied the horses in a shady spot, and started on foot for our objective. But in spite of this early start it was later than we anticipated when we finally reached our destination

65

on the ridgetop—progress had been wearisome and slow because of brush and windfalls.

A black thunder cloud with much sharp lightning was upon us as we completed our work and started back. Our way led through an area of old burn where there were many snags standing, and with the first raindrops came a blinding flash as lightning hit one of these old trees. Quickly we sought shelter under a shelving rock which gave us some protection from the downpour that followed—one of the heaviest showers of the entire season. Lightening struck another snag near where we were crouching with a detonation so violent that it left us dazed. The cloudburst was welcome for it lessened the danger from fire, but the blinding flashes of lightening were rather frightening.

After the rain stopped we came out of hiding and looked around. We saw where the snag had been hit—a deep gash in the bleached veteran and much splintered wood evidenced the force of the bolt. We were thankful to find that no other fires had been set here or in the immediate vicinity, for we had no desire to end our last day corraling blazes. Such tools as we had were back in camp, a couple of hours away.

A double rainbow arched across the sky as we came off the hill; drops of water glistened on the brush in the sunshine. We were well-soaked from the waist down when we finally reached camp. Things there were in a somewhat sodden state, but we threw saddles and packs on the horses and started homeward. Hours later when we reached the station, our clothing was dry and the storm only a memory.

The boys at the station were rather glum as they sat around sipping strong coffee and ruminating about a continuance of the war. We told of our experience in the lightning storm, and they showed us rattles from several large snakes they had killed around the station in our absence.

There seemed no reason for me to stay longer at Bear Creek, and I made plans to leave the next day. I would go back over the trail which I had followed coming in. There was a trail crew camped near the mouth of Cub Creek, and there I could spend the first night on the way. This was a wise plan, and worked it very well.

When I reached Elk Lake near the top of the Bitterroot Divide, I saw that I could not reach Como in time for the afternoon train to Missoula, so slowed my pace and took a shortcut to Darby, where I had a good meal and a comfortable night between sheets. I took the morning train, which reached Missoula about noon. As souvenirs of the trip I was able to show several sets of rattles taken from reptiles in the Selway wilderness!

# A Road to Far Places

The idea of a Forest Service "showboat" which would carry the message of forest conservation to remote parts of the Northern Region originated, I believe, with Theodore Shoemaker soon after he came to Missoula to head up the Office of Public Relations. Calling me to his desk one day he said, "You know, K.D., there are people living in eastern Montana who know littie or nothing of the aims of our forest conservation policy—folks who never even saw trees growing in a forest. We need to reach these people and tell them in terms they can understand just what benefits they will gain from proper management of the forests—their forests. We preach conservation sermons to those living here in the timbered parts of the Region; let's carry the gospel to the kids and grownups in treeless eastern Montana."

And so our "showboat" program was initiated. We had the loan of a new Model T Ford pickup from the car pool. On either side of the body was emblazoned the emblem of the Forest Service—the familiar pine-tree shield. There was ample space for a Homelite generator, a DeVry motion picture projector, a Baloptican for showing slides, a screen, a two-reel motion picture titled "What the Forests Mean to You," and two bedrolls for emergency sleep-outs. All being securely packed and protected with a canvas cover, we started one Sunday morning in early October of 1926 on a tour to fulfill bookings in twenty-eight towns—most of them small places where entertainment such as we could present would be most welcome. Without exception county school superintendents and school principals had been very cooperative in helping us arrange our schedule and giving us advance publicity.

That Sunday evening a goodly number of residents of Judith Gap were on hand in the high school auditorium to see and hear what we had to offer. It was an intelligent audience, and many lingered after the show to tell us of trips they had taken in the Little Belt and Snowy Mountains, and to assure us that they never tossed burning materials from the car, and always, yes *always*, put out their campfire before leaving. This initial meeting set the pattern for evenings to come; we were greatly encouraged by the friendly reception accorded us in Judith Gap. This, by the way, was the only Sunday evening on which we held a meeting.

Monday night saw us in Lewistown, where we had very good attendance in the high school. On the following morning we headed eastward (over the route now followed by Highway 20) to fill engagements in Jordan, Circle, and points beyond. Leaving the mountains behind, we started across a plains country impressive in its vastness—a gently rolling land with here and there a scattered growth of ponderosa pines or an occasional coulee where green ash trees and boxelders were conspicuous in yellow autumn dress. Often horned larks flew up on our approach, always staying a safe distance ahead of the car. Several times we saw flocks of sage hens who didn't seem disturbed as we passed by. There wasn't much of a road in those days; for miles the way consisted of wheel ruts in the sand. In one spot we were stuck in a sanddrift and had to use the shovel to get out. At noon as we basked in the warm Indian Summer sunshine at the base of a butte crowned by a solitary pine tree, we ate a watermelon which we had carried wrapped in a moist gunny sack. No watermelon that I can remember ever tasted better!

Few cars traveled this lonely road in those days, and by the middle of the afternoon we began to wonder where it was leading us. There were no visible habitations—no one of whom we might inquire the way. We wanted to reach Jordan in time for supper, and hoped for a little extra time to set up our equipment for the evening show before sitting down to eat. At that time, except for a Delco plant or two, there was no electric current in Jordan, and we knew we would have to provide our own with the Homelite generator. The sun was getting low when at last we spotted a wagon drawn by a team of horses coming over a swell in the prairie. On drawing closer we saw that the driver was a lanky man dressed in Levis and wearing a tattered mackinaw. His deep-set eyes peered at us from beneath the brim of a western hat, under the frayed ribbon of which were tucked a dozen or more matches. Beside him on the wagon seat was a large collie, apparently well-broken to this means of transportation. From numerous packages and sacks of feed in the wagon box it was evident the man had been to town for supplies and was on the way home. "Yes," he said in answer to our question, "Jordan is over there. You'll see it right away—after you get over the next rise. Not much of a place but it does for a town in these parts."

Thanking him, we pressed hard on the low pedal (remembered by Model T owners) and surmounted the rise. There it was—Jordan, the little town lying in the valley of Big Dry Creek—a welcome oasis for two dusty and sun-burned showmen!

Our coming had been well-advertised. Numerous children took notice of our progress down the main street—the shyest staring at us in silence, the more precocious waving and shouting greetings. Later we were told that some of these children had never seen a motion picture. We felt it a privilege to bring a bit of evening entertainment to that isolated hamlet. It was with a feeling of deep satisfaction that we waved goodbye the next morning as we left for Circle, the next stop on our itinerary.

Circle in those days had no railroad—that would come later. But the community spirit and the enthusiasm of the teaching staff were impressive. Many children from outlying ranches boarded in town during the school season, and a bright bunch of youngsters they were. Several of the teachers made us welcome at their dormitory, a large dwelling house with sleeping quarters on the second floor partitioned off by gingham curtains. Substantial meals were served in the dining room below; I am sure no one went away from the table hungry. At breakfast we met an attractive girl who had ridden twenty miles that morning to be interviewed in connection with a teaching position at the high school. We asked her how the town got its name, and her answer confirmed what we had already guessed: one of the big cattle outfits in the vicinity used a circle for a brand.

Wibaux, once a famous cow town and cattle shipping point, was the place farthest from home reached on our trip. Here on a hillock at the outskirts of the village is a statue of Pierre Wibaux, a cattle baron who is said to have owned 75,000 head of cattle in his prosperous years. In its heyday Wibaux had the largest loading pens in the country, from which animals were shipped to eastern markets. Theodore Roosevelt is known to have driven cattle here from Medora, North Dakota, for shipping.

*Splendid Was the Trail*

The hectic days of the cattle ranges have gone, but when we were in Wibaux old-timers still loved to recall for the benefit of gullible listeners the times when cowhands rode merrily up and down the dusty main street twirling ropes and brandishing six-guns. We were fed well in the little town, and our show drew a good crowd. We hope Pierre's statue will stand there for many years to come as a reminder of bellowing herds and hard-riding cowboys.

Near Glendive is a spectacular bit of badland country, most of which is now included in the Makoshica State Park. Having time to spare on the afternoon preceding an evening show, Shoemaker and I paid a visit to this remarkable spot. It was a sunny afternoon with the stimulating touch of fall in the air—an afternoon perfect for taking pictures. Driving as far as we could up a main wash which leads into the heart of the area, we found ourselves gazing at an array of formations so fantastic as to defy adequate description. Putting on our hiking boots and taking the camera, we scrambled up to a point from which there was a comprehensive view of this amazing jumble of sculptures. One group of pedestals surmounted by flat capstones resembled a big chair, and we dubbed it the Goblin's Armchair.

Roaming through this weird land called for a lot of physical exertion; the clay banks offered little secure footing, but luckily there was some brush and juniper which we could grasp when the going was precarious. Finally we reached the top of the slope and stood on the edge of an expanse of rolling grassland—a plateau into which the badlands were advancing by the irresistible forces of erosion. In places one could see thin coal beds which had been exposed to form horizontal black lines in the light-colored clay beds. When we started back to the pickup the sun was setting; the fall day seemed all too short.

Staying over Sunday in Baker, our next stop after Glendive, gave us a chance to visit the Medicine Rocks, now a state park, some thirty miles south of town on the road to Ekalaka. The term "medicine" was often applied by the Indians to anything strange or unusual. Certainly these rocky sentinels of sandstone, rising to heights of forty or fifty feet above the prairie, deserve a name denoting something out of the ordinary. Standing in the midst of these outlandish figures, one might imagine himself in the abode of trolls or other mythical creatures. How long erosion has been at work forming these oddities one cannot even guess. It must suffice to know that they are here today for us to enjoy.

Theodore Roosevelt paid a visit to these rocks when he was ranching on the Little Missouri, and he described them vividly in his book, The Hunting *Trips of a Ranchman*. He ends his description by saying, "Altogether it was as fantastically beautiful a place as I have ever seen; it seemed impossible that the hand of man should not have had something to do with its formation." To fully appreciate Roosevelt's remark one should stay as we did until the shadows lengthen across the prairie, and the dun-colored rocks take on a rosy glow from the setting sun. At such a time this fascinating spot becomes a land of enchantment.

On the Monday morning after our visit to the Rocks we left Baker in time to keep an appointment that evening in Sidney. This town on the Yellowstone River is notable for its irrigated farm lands which support a considerable sugar beet industry. Because of our early start that day we had a whole afternoon to visit the site of old Fort Union, which was located on the Missouri River near its junction with the Yellowstone. The fort, consisting of log buildings protected by a high stockade, was built about 1828 by the American Fur Company as a center of its trade with the Assiniboine Indians, who often came long distances to trade pelts for supplies, rifles, ammunition, and, alas—strong waters. On June 17, 1832, the Yellowstone, first steamboat to come this far up the Missouri, arrived from St. Louis.

The actual confluence of the Missouri and the Yellowstone is in North Dakota a short distance east of the Montana line. Here the Lewis and Clark party was reunited on August 12, 1806. Lewis with nine men had left Clark and other members of the party at Traveler's Rest, a spot on Lolo Creek near Missoula, to follow up the Blackfoot River, and cross the Continental Divide at the headwaters of that stream. Clark, on the other hand, went up the Bitterroot Valley, crossed the Divide at Gibbon's Pass, and followed the Big Hole, Beaverhead, and Jefferson Rivers to the Missouri. Here he sent a detail to meet Lewis at the mouth of the Marias River, while he himself took an overland course to the Yellowstone.

Shoemaker and I were both eager to stand as close as possible to the meeting place of these two great rivers. Crawling under barbed wire fences and skirting cultivated fields, we finally reached a point where I could set up the camera for a commanding view. There was nothing spectacular about the scene—this view of river bottom lands with fringes of cottonwoods—but because of its historic implications it held great interest for us. As we stood there we thought of the jubilation the Lewis and Clark party must have felt on being together again for the long journey back to civilization. We pictured, too, the canoes and pirogues of Indians and traders who used these waterways when the American Fur Company was in existence—the stern-wheelers working their way up the Missouri to bring prospectors and settlers to a new land—the wild creatures that once roamed the plains and river breaks in countless numbers. When we got back to the pickup we were muddy and tired and had missed our lunch, but we felt amply repaid by having exposed a film at the meeting place of two great waters, the Missouri and the Yellowstone.

On reaching Plentywood a few days later we asked some of the oldtimers how the town got its name for, as far as we could observe, wood was anything but plentiful here. One gray-bearded resident seated close to the stove in the small stuffy room which served as a lobby repeated a legend telling of a cowboy who ran across a cache of wood on the prairie—plenty of wood to cook his bacon and bannocks. Who this lone rider was or who left the store of wood which he discovered no one knows. We wish that the cowboy artist, Charlie Russell, might have used the incident as a basis for one of his pictures!

At Scobey on the day following, we sat at dinner with an official of the Immigration Service who was off duty from his post at the Canadian boundary a few miles to the north. The meal, served by a friendly woman wearing a gingham apron tied on with a big bow in the back, was ample and well-cooked, and the conversation pleasant to remember. The Federal man showed an interest in what we were doing, and came to our program later in the evening.

Poplar, Wolf Point, Glasgow, Malta—these and other places in northern Montana were on our schedule and were visited in due time. Some far-seeing souls may have envisaged the huge Fort Peck dam now built on the Missouri River a few miles south and east of Glasgow, but we heard nothing of it then. At Malta we turned off the main line of travel to visit the Little Rockies, where two of the most colorful Montana mining camps are located.

The Little Rockies might be likened to a mountainous island rising from a surrounding sea of plains. Big Scraggy Peak is the crowning summit, with an elevation of 5896 feet, a conspicuous landmark seen from many directions. It is a comparatively small area, the Little Rockies, but it boasts a niche in Montana history out of all proportion to its size. At one time Zortman, its principal town, had the world's second largest cyanide mill for the recovery of gold, and the nearby Ruby Mine is said to have produced over three and a half million dollars in gold bricks before it was closed down at the end of the last century. The town was named for Pete Zortman, who came to the locality about 1880.

Much lurid history has been written about Landusky, a camp named for "Pike" Landusky, a somewhat lawless character who was notorious as a gun-fighter. He and his partner, Bob Orman, discovered the August Mine in 1893, which proved a rich gold producer. When news of the discovery leaked out other miners and prospectors poured in and Landusky became a thriving town. "Pike" built a saloon for Jew Jake, who had lost a leg in a gun-fight with a deputy sheriff, and this establishment became notorious as a hangout for desperados. Jew Jake delighted his customers by using a Springfield rifle for a crutch, so they say. In listening to tales about early days in Zortman and Landusky, one tries to guess how much the stories have grown in the telling throughout the years.

The Little Rockies comprise a division of the Lewis and Clark National Forest. At the time Shoemaker and I were there, Stacy Eckert was ranger in charge and we spent a pleasant evening in his home hearing about his experiences both before and after going into Government service. He had a lot to say about the rough days in the camps—how on one occasion he had to duck behind a cluster of beer barrels to avoid being hit by stray bullets fired in a fracas at one of the saloons.

In 1947 Stacy and I went back again together to the Little Rockies, chiefly to get pictures which would show the results of a disastrous fire which swept the hills in 1936. On this trip we paid a visit to some Indian pictographs, where with the aid of a flash gun I got some color slides of these drawings done by some primitive artist of long ago. This trip was made in June, and the yucca (probably glauca) was in bloom on the grassy hillsides. From these open slopes there were wide views to the south and west across miles of plains to mountain ranges on the far horizon. This was country where bison herds once ranged; Stacy said he could remember the time when skulls of the great animals could be picked up almost anywhere. We took our time roaming over the hills that afternoon and enjoyed every minute of the hike. I felt fortunate in having Stacy for a guide—a man familiar with the lay of the land and well-versed in the history of the country.

But now to return to the trip with Shoemaker. After leaving Zortman it was our plan to cross the Fort Belknap Indian Reservation to Harlem and thence to Chinook where an evening show was scheduled. The road we followed was primitive, with much to admire and photograph in places, especially in the canyon of Peoples Creek. We stopped at St. Paul's Mission, founded by a Jesuit, Father Eberschweiller, in 1886. Nearby on a hillside was an Indian place of burial, where before the custom was discontinued a few years earlier, bodies were left on the surface of the ground, either in some sort of wrapping or in a rude box or coffin. The box used in one burial of which I took a picture had fallen apart to disclose a body surrounded by trinkets and utensils which had presumably been used by the deceased in his or her lifetime. At another place the bodies of two infants were in an old-fashioned tin trunk, the cover of which was missing. Turning away from this macabre spectacle, we headed across a country of gently rolling grassy hills—a country bathed that day in cheerful October sunshine.

The ride across the Belknap Reservation proved to take longer than we expected, and our concern about getting to Chinook in time for our evening program was heightened when we suspected we had made a wrong turn somewhere. The road we were following degenerated into wheel tracks cut deep in the sod, and we seemed to be headed towards a point too far to the east of our destination. As there was no one to give us directions, we proceeded, trusting that the tracks we were following would not peter out entirely. But shortly a speck appeared in the distance, and this tiny object grew larger and larger as we approached, until we found ourselves talking with an Indian on a very scrawny cayuse. He was not a fluent talker, but he assured us if we kept going we would come out "some place—some where." This place proved to be Harlem, which we reached just at sundown. From there Chinook was a short drive over a good road. A gratifying turnout of children and adults was awaiting our arrival, so we postponed our supper until after the show. This was a day we remembered and talked about for years. The map now shows a paved road on the course we followed across the Fort Belknap Reservation.

On a sunny afternoon, brisk and cold, we reached Fort Benton, our last stop on the trip. There was plenty of time to look about in the historic old town, and to take pictures of the fort and the river bank with its row of cottonwood trees—trees which we felt sure dated back to days when steamboats tied up to the quay to discharge their cargoes. This was the head of navigation; from here travel westward was over the wagon road built by Capt. John Mullan—a way which became famous as the Mullan Road. By today's standards it was rough indeed—this road across the mountains to Walla Walla, near the Columbia River—but for many years it served well the settlements along its course.

Our motion picture machine, a suitcase model which accommodated thousand foot reels of 35 mm film had served us well on the trip. But on this last night at Fort Benton, like the Deacon's One Horse Shay in the well-loved poem, it suddenly went to pieces right in the middle of a tense scene showing what happens when a flipper tosses a cigarette stub out of the car window. It was certainly embarrassing to have a dark screen at that crucial point, and there were howls of disappointment from the kids in the audience. A group of hoodlums, who had started their Saturday night celebration before the show, caused quite a disturbance until they were forced to leave by the town constable. In the meantime we got the slides and Balopticon into action, and Shoemaker wound up the program with one of the most interesting presentations I heard him give on the trip.

Seated in our room at the hotel that night—a room whose plush furnishings dated back to the boom days of river traffic—we summed up the accomplishments of the trip. In the twenty-eight towns we had visited, a total of over five thousand men, women, and children had been given a better understanding of the important role well-managed forests play in the economy of the country. For us the trip had been a happy experience. We had enjoyed the grass-roots hospitality of many little towns off the beaten trail of tourist travel—the bright, sunny days which made it possible to get pictures of unusual places along the way—the chance to learn that far from being a monotonous expanse of uninteresting plains, Montana east of the mountains has a distinctive charm of its own.

# Fire Call!

In the late twenties it became apparent that if the Region was to initiate and build up a comprehensive picture record of our efforts to prevent and control forest fires a photographer would have to be assigned to the job and left free of other duties during the dry months to go where he felt pictures could be obtained. I was chosen as that man; the job was exciting, and presented a real challenge. For many summers I stood ready to leave Missoula on a minute's notice day or night. Often a call would come from

*Splendid Was the Trail*

the Office of Forest Fire control; sometimes from a forest supervisor who saw opportunities for pictures on his forest.

In those days photographers went into the field loaded down with much cumbersome equipment. For many years I took still pictures of fires with a 5 x 7 or a 4 x 5 inch Graflex, and in some cases with a 6 1/2 x 8 1/2 inch view camera which had to be mounted on a tripod. Motion pictures were taken with a Bell and Howell Eyemo on standard 35 millimeter film which came in 100 foot lengths packed in tin boxes which permitted loading in partial daylight. I always carried with me a changing-bag in which I could load cut film magazines or check trouble if anything went wrong with the Eyemo. On more than one occasion I had to change a dozen or more cut films for the view camera in this bag. never kept a record of the many miles I hiked uphill and downhill on the fireline with this weighty paraphernalia carried in a packsack.

There was sure to be plenty of excitement on these fire assignments. There were happenings which made one smile—as when a crew of CCC boys hung their shirts on the brush and lost them to the fire when the wind shifted. More often here was tragedy to observe—especially if one considers tragic the incineration of thousands of acres of fine timber, with the consequent loss of game animals, or perhaps the burning of the home of some backwoods settler. I can laugh when remembering the burning shirts, but all sense of merriment fades when I recall hat awful night when the Halfmoon fire made its big run and brought ruin to he cherished stand of old-growth timber at the lower end of MacDonald Lake in Glacier Park—a bit of forest which had impressed me on my first entry to the Park in 1916.

The summer of 1929 had been extremely dry, and by August the forest fire danger had become explosive. The plant of the Halfmoon Lumber Company north and west of Columbia Falls was the scene of a mean blaze, said to have been started by a trashburner. A crew of millworkers augmented by other help had subdued it to a point where it was felt safe to leave it in the care of a few men who would watch for any stray sparks which might jump the trenches. This was a sad mistake in strategy; a wind came up in the night, and before extra help could arrive the blaze had jumped the lines and was again roaring out of control, crowning into the tree-tops where the wind snatched flaming brands to spread fire far and wide in the tinder-dry woods.

The Missoula Office of Fire Control kept in touch with the situation and about the middle of the afternoon of the following day decided to send me to photograph the disaster. Nothing could have pleased me more. Within an hour I was on the road in a Model A Ford loaded with camera equipment, a bedroll, and a few emergency rations. A parting word of advice from G. I. Porter is still remembered: "Don't let them put you to work, K.D., keeping time or doing other menial jobs. Your assignment is to travel around and get pictures." What more could a photographer ask?

It was dark when I reached Columbia Falls, and after supper I unrolled my sleeping bag in the municipal tourist park. There were many sleepers there that night, presumably men off the fire line. Much smoke was drifting in from the fire which at that time was burning on the hills to the northward. The wind had abated somewhat in the early evening, and desperate attempts were being made to bring the blaze under control; with the coming of daylight it appeared that the all-night struggle had been successful. But before ten o'clock the wind came up again and the flames went on another rampage. That proved to be a day I shall never forget—likewise the ghastly night which followed.

Throughout the forenoon smokeclouds billowed skyward from Teakettle Mountain—an impressive sight. I got as close as possible with my car and took stills and movies of the spectacle. Then by way of the North Fork road and much footwork I paid a visit to the fire lines that had been built during the fight—lines that were now of no use whatsoever. Morale was at low ebb among the tired men who were seeing their efforts of the night before brought to naught by the wind which was blowing harder and harder as the day advanced. They were a good bunch of men—none better ever trenched a fire. In the crew were loggers, mill-workers, ranchers, foresters—men who were used to hard labor and not easily discouraged. Fires were trenched in those lays by men wielding hand tools. A mechanical trencher was being developed, but it came too late to be used on this fire. Pacific Pumpers were coming into use, and were very effective in dampening a blaze where water was available.

In late afternoon I went over to the Coram Ranger Station, and after a talk with the men assembled there I decided to make it my temporary headquarters. Meals at the cookhouse were excellent, and there were plenty of spots to spread a sleeping bag. A number of District Office men had gathered there to help in a supervisory capacity. Regional Forester Evan Kelley, affectionately referred to as "The Major," had come up from Missoula to help in strategic planning. Harry Gisborne, fire research specialist from the Rocky Mountain Experiment Station, was observing events from nearby Desert Mountain Lookout. His story of what transpired that night appeared in the magazine American Forests, and was reprinted in the Reader's Digest. Howard Flint, quiet and unassuming as usual, was observing and making notes of the behavior of the fire—notes which proved invaluable for a final report prepared later. After supper he suggested that we take my car and do a little reconnaissance of the fire-front. This was about seven o'clock—not late but much darker than usual at that time on account of the smoke pall. We could see a great smoke cloud churning slowly high above Teakettle Mountain. It was such an impressive sight that I set up the view camera and took a couple of exposures; the resulting pictures were among the best I was able to get of going fires that summer.

Not far from Coram we took a rough road which led to the Great Northern Railroad at the whistlestop known as Egan, a loading point for ties. Tiemakers and loggers had left much slash scattered on the ground in this locality, and Howard remarked that the whole country thereabouts was one big firetrap. The truth of this remark was demonstrated within minutes. No

*Splendid Was the Trail*

sooner had we reached the rails than we discovered several spot fires of alarming size across the track, and felt hot ashes blowing our way. "Let's go," said Howard - and we did. Before we could turn the car around, a flying brand had started a spot fire at the side of the road over which we had come and was spreading with explosive rapidity. Never before had I seen fire travel so fast, even in the driest woods. Before we reached the main highway the slash-filled forest was a flaming inferno.

Back at the ranger station we got a report from Gisborne that undoubtedly the fire was spreading towards Glacier Park. Later the flames raced up Desert Mountain, but stopped short of the lookout. Gisborne had a few bad moments which he describes vividly in his story.

Next morning the whole country was shrouded in yellow smog. On either side of the road to Belton where the day before had been a green forest was a smoking ruin—a blackened waste where here and there stumps and snags were still holding fire: now and then an old veteran whose base had been eaten away by the flames would topple over to land among flying ashes with a resounding thud. Soon we passed a family coming down the road with their household goods in a wagon, and with a couple of boys leading milk cows.

"Yes, we're burned out," said the mother, who was comforting several children. Dazed and aimless they seemed, as they answered our questions. "Everything's gone," said the woman. "Everything except what's in the wagon. We'll camp in the park at Columbia Falls. There'll be others there too." Later we passed the hot ruins of the house a few steps off the road beyond Lake Five.

We found the fire had taken toll of some of the finest timber in the lower MacDonald Creek Valley and had left enduring scars in Glacier Park. Here at the southern end of MacDonald Lake—the sacred dancing lake of the Kootenai Indians—had been a magnificent stand of old-growth western whitepine, western redcedar, and larch. Now only the scorched boles remained. The flames had raced up the slopes of the Belton Hills, but the extent of the damage was not apparent because of smoke. Park rangers stood around with glum faces, eyes reddened from their all-night vigil in the smoke. The whine of Pacific Pumpers wetting down the hot-spots made conversation difficult.

Depressing indeed was the thought that I am sure came to most of us on that gloomy morning. "This fire was man-caused; this agony could have been prevented!"

On my return to Missoula a day or so later I was asked to go over to the Lochsa country where another big fire, the Bald Mountain blaze, was burning uncontrolled. It was of gigantic size and spreading rapidly in spite of all efforts being made to control it. And so after a day spent in the darkroom developing an accumulation of exposed film, I took off for the Powell Ranger Station which was the center of activities connected with this exciting campaign.

This was on Ed MacKay's district, and Ed surely had things moving.

MacKay's reputation as a man of exceptional efficiency and stamina in an emergency was well-earned, and here on the Powell District at this time I had ample opportunity to observe those qualities which brought him the high regard of all who ever knew or worked with him.

Ed had placed Jack Clack in charge of keeping supplies moving to the camps down river, a job of no mean size. In addition he kept an eye on activities—the unloading of supply trucks, the dispatching of fresh contingents of firefighters to the camps where they were needed, answering questions in connection with time slips—giving ear to the many petty details that always arise in connection with a fire organization. Jack, notably undisturbed and cool, had the situation well in hand, leaving Ed to spend a good deal of time on the fireline.

My mainstay in equipment on this trip was a 5 x 7 Graflex, considered a superb camera in those days, but bulky and heavy to carry. My DeVry movie camera was to be left behind on this trip. Most of the action at the moment was down river some twelve miles, where a camp had been established at the Jerry Johnson cabin.

"No need to walk," said Jack after taking a look at my impedimenta. "Save your strength for climbing the hills." Without more words he took me over to the corral, picked a gentle horse for me, and devised a sling from a gunny sack whereby I could carry the Graflex on the saddle horn.

"See you when you get back," was Jack's terse remark as I rode out of camp.

Arriving at Jerry Johnson I checked in with the camp clerk and picked out a kapok from a pile of firefighters' beds, as I intended to spend the night there. Then with the Graflex in a borrowed packsack I started up the hill towards the fireline.

I met Ed high on the ridge; there he took charge of the Graflex—a big relief for me. The afternoon was hot, and so were we. I think Ed as well as myself welcomed an excuse to stop and talk for a few minutes, and I told him the latest news of the fire situation in other parts of the Region. The fireline was not far away; as we talked whiffs of smoke were wafted to us by an almost imperceptible current of parched, stifling air. It was truly fire weather—hot, dry, explosive. Perhaps Ed more than I realized the critical nature of the situation at the moment, but never in the course of the conversation did he voice discouragement or hint that we would not be able to get the better of the blaze. Walking to a more open spot on the ridge we stood looking across the valley to another smoke when a boy came running down the slope to the spot where we were standing to tell us a man had been killed by a falling snag. Here Ed and I parted company, he to climb up to the scene of the accident, I to go back down to camp to await developments.

I had no desire to take more pictures that afternoon as I took the camera from Ed.

The way down to the river was steep and arduous. I could hear camp sounds far below as I descended. Preparations for supper were in progress, and soon the men would be coming in for the meal, served buffet style where each man helped himself and then looked for a spot where he could balance his tin plate on his knees. What a surprise all this would be to Jerry Johnson, I thought, could he come back to his old spot in the wilderness. Jerry was a prospector who came into the Lochsa country to search for a lost gold lead described to him by an Indian known as "Indian Issac." Issac, who spoke only a few words of broken English, had difficulty in describing the exact location of the find. Unfortunately, as usually happens in such cases, the only one having the needed information which would lead to fabulous riches—in this case Indian Issac—died before his mission was accomplished. But they say Jerry Johnson never abandoned the search.

Before dark most of the men were in from the line, and after eating broke up in small groups to talk together in subdued voices of the tragic event which had taken place that afternoon. Several of the men asked for their time, and it took considerable talking to persuade them to remain on the job. Lastly the little party bearing the body of the victim on an improvised stretcher reached camp; it had been no easy task for the bearers to come down the mountainside with their grim burden, and one and all gave a sigh of relief as they set it down in the rear of the office tent.

An ambulance was ordered to come to Powell, and plans were made to transport the body there on a mule. The grim task of wrapping it in blankets and canvas was soon accomplished, and the bundle was lashed in a more or less upright position to one side of a decker packsaddle mounted on a gentle mule who stood quietly during the operation. To balance the load somewhat a bundle was tied to the opposite side of the saddle.

But then an unforeseen problem presented itself. The packer refused to go to Powell without an escort. To his way of thinking, no man should travel alone through the dark woods with such a load. "Twelve miles with a dead man? No sir," said he.

Inasmuch as I was planning to ride back to Powell in the morning anyway, and as no one else seemed hankering for the job, I agreed to boost the morale of the packer by bringing up the rear of the little procession.

The woods were dark that night, and I wondered how the animals could find the trail. About all that I could see from my position in the rear was the long white bundle which bobbed up and down as if keeping time to the rhythmic gait of the mule. There were few sounds; only the thud of hoof-beats and the background murmur of the Lochsa broke the stillness. I was tired enough to relax in the saddle, and I enjoyed the quietness which seemed like a benediction to a long, strenuous day.

Anticipating our arrival, Jack Clack was waiting to supervise the transfer of our cargo to the ambulance. I spread my bed in Jack's tent and turned in for a few hours sleep; the packer did the same after taking the animals over to the corral for a feed of oats, and I did not see him again, for he was on his way back to Jerry Johnson before I was up next morning.

A few miles down stream from Powell there was, in the days before the road was built, a spot where the trail made a sharp turn around a rocky point—an ideal place to get pictures of packstrings on the move. Jack and I hiked down to this spot the following afternoon, and there I got the most usable pictures I took on the trip. But the picture which I would have prized above all others was unobtainable—a shot of a white bundle on a mule moving through the dark, silent woods. That picture has been left to memory.

The summer of 1934 provided more opportunities for pictures of spectacular fires in Region One. At this time the Civilian Conservation Corps program was in full swing and I was asked to get as many pictures as possible which would show how the boys were furthering our fire control efforts. In addition to still shots, movies were needed which would show lively action on the line, with plenty of smoke and flame as a backdrop. My experience on the Half Moon fire would be of tremendous help in getting such pictures and I needed no urging to get my equipment in readiness for a quick get-away. Most of the time for stills I was now using the 4 x 5 Graflex with cut film magazines, which were often loaded in the changing bag; two filters, the K2 yellow and the Orange A, I found indispensable, especially for photographing smoke clouds. For most scenes, especially where there was much contrast such as one encounters in the woods, I preferred Eastman's Portrait Panchromatic film—not a fast film but an emulsion which seemed to fit my needs very nicely. For the 35 mm. Bell and Howell Eyemo (recently acquired) equipped with three lenses in a turret mounting, I used a standard emulsion which came packed in 100-foot magazines.

A call to action came about three o' clock in the afternoon of a blazing hot August day. A serious blaze was reported to be out of control near Selway Falls on the Selway River in Idaho; crews of CCC boys were being rushed to the scene, and the situation gave promise of furnishing rare picture opportunities. I was to take the next plane to Spokane, leaving in about an hour; there I would be furnished with a car from the pool at the Forest Service warehouse. And in no time at all I found myself at the Missoula Airport, then just beyond the fairgrounds, with two packsacks of camera equipment, a bedroll, a dufflebag of personal necessities, and the blessing of O.C. Bradeen who had provided me with transportation from town. In those days Northwest Airlines used Ford Tri-motors and the flight to Spokane took over an hour.

I reached Selway Falls in time for breakfast. One contingent of boys had been on the job since daylight; other groups were arriving at intervals, having spent most of the night on the road. Tired and dusty, they climbed down from the trucks to revive their drooping spirits with a warm meal. Some of the boys had been furnished with smokechaser rations, which they ate enroute during the night. Remarks about this unpalatable but said-to-be nourishing type of grub were mostly uncomplimentary, and tended to arouse sympathy for the poor smokechaser.

A lad with a New York City background was assigned to act as guide and help me carry my equipment. He was a willing helper, but I could see life in the woods was a new experience for him. His buddies on the fireline threw out taunting remarks when they saw him laboring under a load of paraphernalia, a flunky for one of the top brass. I eased the situation somewhat by explaining to a group of interested boys the working of the movie camera and getting them to do a little fast action for me on a hot-spot flaring up conveniently near at hand.

The explosive nature of the forest fuels became apparent as the day progressed. Any slight breeze would fan a smoldering spark into a serious blaze within a few seconds. Late in the afternoon thunderheads built up to the south and the country was bombarded by a storm which furnished plenty of lightning but only a sprinkle of rain—a situation much dreaded by the forest firefighter. That evening lookout stations reported many lightning strikes on the higher ridges.

The next morning as I was getting pictures of the breakfast grub line, the camp boss beckoned me aside to say there was a big smoke cloud rising above MacLendon Butte, a few miles to the north.

"It looks like a whopper," said he, "you can get close on a road, and if you leave right away there is a fine chance for pictures. Better get going."

Going down the Selway River to its junction with the Lochsa I turned up the latter stream to a road which led to Van Camp Lookout and beyond. In the vicinity of Middle Butte I found a point from which there was an unobstructed view of the great mushrooming cloud which was rising high in the air over McLendon Butte. This was a brushy country which seemed to provide plenty of fuel for the spreading fire. It was a fascinating spectacle, this great cauliflower in the sky, its upper convolutions brilliantly white in the sun, and the whole mass changing shape from moment to moment as one watched. I had no means of making an accurate estimate of the height of the cloud, but a rough guess would place the top about thirty thousand feet above the earth. Every now and then flame and blacker smoke would rocket up from ground level as some pitchy snag was ignited or fire got into a patch of highly inflammable green timber. A slight drift of air from the south carried the smoke away to the north and left the atmosphere between me and the cloud clear for picture taking. I got very satisfactory views at intervals throughout the afternoon—especially movie shots taken at four and eight frames per second to accentuate the churning and billowing of the smoke mass.

As the afternoon advanced thunderheads built up in the southwest, and towards evening lightning could be seen in many directions. Snags on Coolwater Ridge directly to the south seemed to provide targets for frequent strikes, but not a drop of rain reached the ground. Curtains of rain hung from the clouds, but any moisture they may have held was snatched up and evaporated quickly by air rising from the heated earth. Far in the southwest sheet lightning flickered continuously, making a display which became more spectacular with the coming of evening darkness. The rumble of distant thunder was continuous—but still no rain.

I decided to spend the night on Middle Butte. Here was a decrepit lookout structure and a rustic bunk on which I could spread my bed; supper would consist of smokechaser rations, prepared with water from a five-gallon tin. It seemed a carefree life, that evening on Middle Butte, and I congratulated myself on having such a choice experience. Frankly I was tired, and my bed looked inviting as I unsnapped the cover of the sleeping bag. My last recollection before falling asleep was the pungent odor of the ceanothus brush, whose glossy leaves had absorbed the warm sunshine of the long summer day.

It must have been about four o'clock the next morning that I became conscious of a sound which seemed to carry me back to the Pryor Mountains where I once made a movie of sheep range management. Startled, I raised up to find myself in the midst of a moving mass of sheep—a band of possibly fifteen hundred animals. Urged on by the insistent barking of two dogs, they bumped against the frame of my bunk, making me think of the saying of old salts, "Shiver my timbers." A yell from me promoted a near-panic of the animals, and caused a serious tie-up in the smooth flow of traffic. The situation was not helped by the sharp yelping of the dogs who did their best to keep the band moving.

Soon the herder was able to make his way to my rustic couch to tell me the fire was spreading all over the country, and that he was getting out while there was time. His camp equipment and supplies were carried on a packhorse he was leading. Finding I had plenty of water, he tied the animal to a post of the look-out and proceeded to get breakfast, which we shared. The bannocks were burned on the bottom, and the coffee (which I did not drink) was boiled until I was sure the last modicum of bitterness was extracted. But it was a friendly moment there on the hilltop, with the morning sun flooding the expanse of ceanothus brush and giving promise of another hot day. We talked of many things—sheep ranges in many parts of the country, Forest Service range practices, the training of sheep dogs. The conversation lacked an intellectual flavor, but it left us with a kindly feeling towards each other. Meanwhile the sheep had moved off the butte and were shaded up wherever they could find spots out of the hot sun. I left shortly and never saw them again.

In a few days the fire had spread over thousands of acres and was being fought on many fronts. Man's efforts to subdue a blaze of this size seemed puny indeed, and had it not been for the excitement of the campaign a sense of frustration and discouragement might have crept in to dampen the ardor of the men involved. But in spite of long hours without much sleep, reports that several spot fires had developed into major conflagrations, the news from the weatherman that we could look for more dry lightning storms—in spite of all this—morale remained high among the men. CCC boys who had hardly been off the pavement before took an awful beating—so they thought—but most of them stood up manfully to the job. Opportunities for pictures seemed unlimited: I soon ran out of movie film and was obliged to place a telephone order for more to be sent from the

Regional Office. In the meantime the Graflex was kept busy getting stills which later would prove useful in telling the story of the fire.

One forenoon while getting gas at the Pete King Ranger Station I met Elers Koch who had just arrived from Missoula to size up the situation. He had caught a ride from Kooskia to Pete King, but was then without transportation, so I invited him to join forces with me. Throwing his bed and duffle in the back seat of the Ford we were soon on our way.

Visiting fire camps which could be reached by car took most of the day. That night we decided to stay at the Van Camp Lookout, which commanded a comprehensive view of the surrounding country—most of which appeared to be on fire. We sat up throughout the hours of darkness watching it burn, and only turned in for a few hours of sleep with the coming of daylight. I have a mental picture of Elers sitting beside the stove quaffing strong coffee and knocking the ashes from his pipe into the ashpan. Through the lookout windows was a panorama of flaming forests, some in the distance, some nearer at hand—a scene such as I had never witnessed before, and have never seen since. We felt safe in our aerie on the mountaintop, convinced that no fire would reach us there.

This fire, which proved to be one of the most extensive in the Region's history, was not controlled completely until the coming of the fall rains and snow: as late as November there were reports of stumps and logs still holding fire. However, the loss of valuable timber was not as great as might have been expected, as much of the country burned had been swept by fire before and was coming up to brush. I heard more than one forester remark that aside from some merchantable timber which was destroyed the fire was a blessing in disguise, as it left the country in much better shape for a new crop of trees by removing an incipient fire hazard. Today (1968) one can see some of the marks left by the fire as he travels through the lower Lochsa country on the Lewis and Clark Highway.

When I felt I had obtained a fair number of usable pictures I drove back to Spokane, turned the car in at the warehouse and took the night train to Missoula. Pullman sheets seemed quite a luxury after nights spent in a kapok sleeping bag. I am sure the porter had me sized up as a character when I insisted on taking a dufflebag bulging with camera equipment and precious exposed film into my berth, and asked him to raise the window as far as it would go. The coal-burning engines of those days produced plenty of cinders on the heavy grades, which were not sifted out to any extent by the window screens.

The early thirties saw the airplane becoming a more important factor in the detection and control of forest fires. This was an era in which through research and experiment we were making rapid strides in the development of new techniques to meet the challenge of the Red Enemy. Air delivery of supplies and equipment to remote areas hitherto reached only by trail was fast becoming a common practice: construction on seven backcountry landing fields was started in 1930, and in the same year free-fall cargo dropping was successfully employed on fires in the St. Joe and Nezperce Forests. The golden days of the smokejumper were fast approaching, when it would be possible to land men by parachute to extinguish blazes before they became conflagrations. From the chrysalis of the sweating smokechaser with his pack and tools was emerging the bright figure of the smokejumper borne to earth by a blossoming canopy of silk.

More innovations, hardly envisioned at the time, would come in the next decade or two: the bombing of fires from the air with chemical retardants, the increased use of helicopters for emergency landings, the cooperation of the Weather Bureau in installing radar equipment to make possible a more accurate forecasting of lightning storms, the building of a fabulous laboratory at Missoula to aid in the scientific study of fire behavior—to mention a few of the forward steps taken during this period.

My first ride in a plane came when Bob Johnson took me with him on a flight to Big Prairie with a load of supplies. There were probably no pilots at that time anywhere with more experience in mountain flying than the two Johnson boys, Bob and Dick, partners who founded the Johnson Flying Service, which has done a vast amount of flying for the Forest Service in the ensuing years. To these intrepid pilots and their assistants goes a lot of credit for the success of our smokejumper program. Both men showed amazing skill in taking their ships in and out of tight places where less experienced pilots would say it couldn't be done. At this writing Bob is still with us; Dick was killed in a crash in the Jackson Hole country in 1945.

On this particular trip the plane, a Travelair, had been stripped of interior furnishings to make room for the cargo. As I recall we put down first at Blanchard Flats to leave a supply of gas which, if needed, could be used to re-fuel on the way back. I sat in the co-pilot's seat beside Bob, and tried to hear above the roar of the engine his explanation of how the updraught of warm air could be utilized to carry the plane over the high ridges. I carried an old 4 x 5 Century camera fitted with a pistol grip and a cut film magazine, but the vibration of the heavily loaded plane made it difficult to get satisfactory pictures. Bob, with consummate skill, made a gentle landing on a field just below the ranger station—a field quite typical of backcountry strips, which usually had plenty of bumps to let one know he was on the ground.

The boys at the station had watched us come out of the sky, and soon a string of mules appeared to pick up our cargo. Here was a wonderful picture chance—the old and the new, mules and an airplane—in the heart of the primitive back-country later to be known as the Bob Marshall Wilderness. The picture was used with the caption, "A New Day Dawns."

It was not until 1940 that the training of smokejumpers really got underway in Region One. The records show that the first actual fire jumps were made on July 12 of that year, when Rufus Robinson of Kooskia, Idaho, and Earl Cooley of Hamilton, Montana, bailed out to a fire on Martin Creek, in the Nezperce Forest. Possibly no activity initiated by the Forest Service has been given such nation-wide publicity as this bold attempt to control fires by aerial firefighters; the demand for pictures was tremendous,

*Splendid Was the Trail*

and I was called on to do my best to obtain them. Both Bob and Dick Johnson cooperated to the fullest extent in helping me get the desired shots. So did Frank Derry, who played an important part in training the first cadres of jumpers. It was he who devised the slotted chute which gave greater maneuverability on the descent.

Blanchard Flats, just north of Clearwater in the Blackfoot Valley, was the site of some of the first practice jumps, and here I spent several forenoons in getting shots of the blossoming canopies as they floated to earth. I found a newly acquired Leica an ideal camera for this work: with a red fitter to darken the sky I obtained many very satisfactory pictures which were widely used. At this time the conventional rip cord was being used to open the chute—a man would count a specified number of seconds and then pull the handle which would release the canopy and allow it to open. There were always a few breathtaking moments as we watched to see this occur. Once a trainee failed to react as he should have done and his chute barely opened in time to prevent a bad landing. He was screened out and never allowed to jump again. In addition to the regular parachute which was worn between the shoulders, an emergency chute was strapped in front at the waist, to be used in case anything went wrong with the main canopy—which seldom, if ever, occurred. I do not recall ever seeing an emergency chute used. The first jumps that I photographed were made from Travelair planes.

Sleeman Creek was used as a training ground at a later time, and I spent several forenoons chasing jumpers coming out of the sky, some to hang up in trees or to make featherbed landings in clumps of young growth. The excitement of those hours is not forgotten. I am sure the jumpers felt, too, that they were getting a full measure of thrill from their training experience. One incident I well remember. With camera in readiness I raced towards a spot where a jumper, so I thought, was about to land. I need not have hurried, as the man never reached the ground as intended: his canopy caught on the crown of a large fir tree, and left him to reach terra firma as best he could. Each jumper carries in a pocket on one leg of his jumpsuit a long rope for just such emergencies. Usually this rope will enable a man to get himself out of a predicament such as this—he can attach one end of the line to a tree limb and lower himself easily to footing below. But in this case the rope was never used: before it could be taken from the pocket a slight breeze caused the chute to slide off the tree crown. We held our breath, for it looked as if the jumper was in for a nasty fall; a thirty-foot drop would certainly have caused grave injury, in spite of the heavily padded jumpsuit the man was wearing. But when calamity seemed inevitable the unexpected again happened. While the man was still clear of the ground, a shroud line caught on a projecting limb of the old fir, leaving the victim dangling within easy reach of the boys who had rushed to help.

Did I get a picture of the incident? I did not. Hunters call it "buck fever;" photographers have no established name for it!

At a later time I spent a forenoon in getting pictures of jumpers as they left the plane. This proved to be a hard assignment. I was given the use of a second Travelair with door removed piloted by Bob Johnson. Bob was adept at drawing close to the plane carrying the jumpers, and cutting the motor at the instant I took a shot, a procedure which lessens the vibration which ruins a picture. Several times Bob maneuvered the plane so as to give me a chance for a shot of a white chute against a timber background. My prize shot shows a jumper just as he has left the plane; the rip cord has been pulled, the pilot chute is open, and the main canopy is about to blossom. (At this time the static line was not in general use—a device which opens the chute automatically, relieving the jumper of any worry in connection with pulling the rip cord.)

It became very apparent during my association with the smokejumpers that the boys were in the game for the thrill of it, and considered the few moments when they were floating to earth ample compensation for the hard work entailed in putting out the fire on the ground. Several times I rode with a jump-crew in a Ford Tri-motor piloted by Dick Johnson to get pictures showing the use of the static line: never once did I detect the least hesitation on the part of any jumper as he stepped into space through the open door of the plane. A grandson tried it for a season—and loved it. "But," said he, "I only jumped on three fires; the rest of the time they had me painting fences!"

# Wilderness Safaris

In the summer of 1916 the Milwaukee Road asked the Forest Service for information about possible tourist trips into the wildlands near Missoula, suggesting they might send a representative from Seattle to look the ground over and get first-hand information for publicity material. Undoubtedly the railroad management was impressed by the lucrative travel business the Great Northern had built up in connection with Glacier Park.

The idea of bringing visitors to the national forests of western Montana was pleasing to our personnel. A cordial invitation was extended to the railroad's representative, with the promise that we would do everything possible to give him a real taste of wilderness adventure. It was decided that a trip into the South Fork of the Flathead country—a region now largely embraced by the Bob Marshall Wilderness—would be most productive of results. Ezra Shaw, ranger on the Seeley Lake District of the old Missoula Forest, was delegated to engineer the trip, and I was to be the photographer for the expedition. Ezra would handle the livestock and do the packing, and I would take pictures and help with the camp chores. In those days I was often called on to function as a camp cook.

Shaw asked me to hire four horses properly equipped and bring them to the Seeley Lake station; also to purchase supplies and gather up needed camp equipment at the Forest Service warehouse, all of which would be shipped to the station by the Seeley Lake mail stage. In the meantime, word confirming the early arrival of our guest was received. A day or two later I was on my way up the Blackfoot Valley, riding a good horse and leading three—two with pack saddles, the other with a riding saddle for our guest.

It was a broiling hot August afternoon when I left Missoula. The rough unpaved road sent up clouds of choking dust, but this didn't smother my enthusiasm for a trip which was taking me away from the office and into the hills. MacNamara's, an inn about sixteen miles from Missoula, was reached just before a terrific thunder storm ended the dust menace. In those days meals, lodging, and horsefeed could be had at this lumberjack hostel, and there I put up for the night. After looking over the questionable sleeping accommodations, I spread my bed in the stable loft where I could listen to the homey noises of the horses below.

The air was sparkling clear the next morning after the rain; there was no dust, and I enjoyed every minute of the ride. That night I stayed at a ranch on the Clearwater River not far from its junction with the Blackfoot. Early in the afternoon of the following day I reached the Seeley Lake station, where I was greeted by Shaw, who had just received a telephone message saying our guest was on the way. Sure enough, in a very short time the mail stage lumbered into the driveway with a lone passenger in the seat with the driver. It was a tiresome trip from Missoula over the unpaved roads of those days, and our guest seemed a little stiff and sore as he climbed down from his perch. His name was Grindell. It would be hard, if not impossible, to find a more likable and cooperative individual. Shaw and I uncrossed our fingers after he had introduced himself—an introduction which was the beginning of a most harmonious trip and lasting friendship.

A bright summer morning set the stage for our horseback trip up the Clearwater Valley and over the divide to headwaters of the Swan River, a ride which took us by a necklace of sequestered lakes—Inez, Alva, Rainy, and Summit. Our guest was new to the saddle, but he seemed to have no trouble in adjusting himself to this unfamiliar mode of transportation. We

*Splendid Was the Trail*

made our first stop at a choice spot on Lake Alva, where a family of Missoula people (I think the name was Bolles) had pitched camp for a week's stay. The long trip from town had been made with a wagon pulled by a team of horses—a journey of several days. There were no tables, fireplaces, or other conveniences there in those days, but their camp was spic and span, and the tents under the towering larches made an attractive picture. A paved highway now encroaches on the lake at this spot—a painful reminder of the frailty of wilderness!

Haying was in full swing at the Gordon Ranch, and we were invited to eat our supper and breakfast with the hayhands, and make use of the corral for our stock. To fill in the time before the evening meal, we rode over to Holland Lake.

The view of this mountain tarn nestled close to the base of the abrupt Swan Range, is remembered as a satisfying climax to the first day of our trip. The air was still and the lake formed a perfect mirror for the lofty peaks to the east. The whole scene was enhanced by the mellow afternoon light. As we listened we could hear the murmur of the falls on Holland Creek just above the inlet of the lake. I became quite familiar with this beautiful lake in the ensuing years, but this first view still remains an impressive memory.

Several years later Mrs. Swan and I had the pleasant experience of taking Robert Sterling Yard and Mrs. Yard to this same spot. Mr. Yard, as many of you will recall, was a militant conservationist who spearheaded the drive to have our national parks preserved inviolate from selfish interests, and gave loyal support to Robert Marshall in the establishment of the Wilderness Society.

Mr. Yard was in Missoula briefly on business connected with the Society, and expressed a desire to see some of our wildlands before continuing his journey. When I was asked by my chief, Richard Hammatt, to plan a trip and act as escort for the party, I immediately thought of Holland Lake as a spot feasible to visit in the available time.

When we reached the lakeshore, Mr. Yard walked alone to the edge of the water and, taking off his hat, stood for some time gazing at the mountains and their reflection in the quiet water. Coming back to the car he said, almost in the manner of one who has had a moving spiritual experience, "You say it's wilderness beyond those mountains?" Then after a pause he added. "I've been very close today!"

There would be much to write should I attempt to tell all that happened on our South Fork trip. There would be word pictures of timberline lakes bordered by meadows still gay with late-blooming flowers; mention of streams which would bring joy to the heart of the angler; tales of encounters with game along the way. Perhaps the crowning highlight would be a description of Big Salmon Lake, a favorite spot with the Indians who came there to catch and smoke their fish. Here we camped for three days before starting home by way of Hahn Creek and Monture Pass. At the Monture Ranger Station we parted with our guest who was taken to Missoula to catch an afternoon train for Seattle. In leaving us he made a remark which I still remember: "I know now what they mean when they talk about God's country!"

In 1924 the Northern Pacific Railroad and the Forest Service organized a party to explore and photograph much the same country we visited with Grindell. Word was getting around rapidly about the wonderful possibilities of that region for wilderness adventure, and dude outfitters were stepping up their efforts to bring people to the country for camping, fishing, and hunting trips. Apparently, the Northern Pacific folks felt justified in making a considerable outlay to finance this trip, as they were not at all stingy with the funds. In the party were Asahel Curtis, well-known mountain photographer, Olin D. Wheeler, a writer who had a two-volume book on the Lewis and Clark Party to his credit, and two motion picture men from the Twin Cities. Theodore Shoemaker, chief of the Office of Public Relations, my boss, was given the job of planning the trip and making the necessary arrangements for transportation and subsistence. Jack Clack was named as his assistant, and a very able one he proved to be. I was to take over the office while Shoemaker was away.

A tentative plan of taking the Northern Pacific party to Big Salmon Lake over the same route followed by Shaw, Grindell and myself was never carried out. As the cavalcade worked its way up the switchbacks above Holland Lake, Curtis was fascinated by glimpses of the Mission Range to the westward where a glacier of considerable size and numerous lingering snowfields were gleaming white in the morning sun. "That's where the pictures are," said Curtis, and insisted that the itinerary be changed accordingly. Jack, who found in Curtis a kindred spirit from the first, agreed that was the thing to do, and after some discussion with Shoemaker the party about-faced and headed across the Swan Valley for Elbow Lake (now Lindbergh Lake) and Glacier Creek. Later events proved the decision to have been a wise one. Curtis got about a hundred remarkably fine 8 x 10 negatives, which remained the property of the Northern Pacific Railroad. He and Jack would leave camp early each morning to spend the day hiking over the trailless high country and picking out choice spots for pictures. Jack told of helping Curtis change films at night under a blanket spread over a smooth ledge at Lagoon Lake, site of the base camp. It was a disappointment to me to remain in town during times like these, but my turn came later!

That summer Shoemaker returned in August with a party of Montana Mountaineers to camp at Lagoon Lake, the logical place for a base camp. The party stayed there for ten days exploring the area, climbing many of the peaks, and giving names to many topographic features. Credit goes to Curtis, according to Jack, for naming Daughter-of-the-Sun Mountain, Turquoise Lake, and Sunrise Glacier; the Mountaineers named Mountaineer Peak, which they were the first organized party to climb. Mrs. Swan, then Miss Ruth Barrows, was a member of this group. Other names given by this party seem very appropriate: Grey Wolf Peak, Lake of the Stars, Panoramic

Peak, Lone Tree Pass, Iceflow Lake among others. It is fitting that a peak in this fine cluster of mountains now bears the name Mt. Shoemaker in honor of the intrepid leader who gave unselfishly of his time to bring people to enjoy this outstanding country. The name has the full approval of the National Board of Geographic Names.

To one not familiar with the Mission Mountains the topography seems a bit confusing. Jack found some of his questions about the terrain unanswered, and wanted to do a little exploring on his own account, wishing principally to follow down Post Creek from its source in Iceflow Lake. He suggested that the two of us go on a back-pack trip—an idea which met with my immediate approval. This trip was my introduction to the Mission Mountains Wilderness, and what an introduction it proved to be!

It was part of the unwritten code of the Forest Service in those days that vacations be taken only after all danger from forest fires was over, usually some time in October. Weather conditions are more or less uncertain at this time; sometimes there is a considerable period of crisp Indian summer weather and quite often a fall storm makes a hiking trip in the hills a rather miserable experience. Jack was an optimist, and we made plans to leave about the middle of the month.

Our outfit included a pup tent, a double-bitted axe, cooking utensils, down quilts, and an ample supply of grub. Jack's pack weighed 65 pounds, mine about 35. Mrs. Clack drove us to Holland Lake one afternoon and there left us with the understanding that we were to call on the telephone when we came out of the hills and tell her where to meet us. Gone now is the log ranger cabin where we spent the night, demolished to make room for more modern improvements, but in those days it was considered a comfortable stopping place.

We were on the trail early the next day, a bright day which we hoped indicated a continuing spell of Indian summer weather. Our way took us across the Swan Valley to Elbow (now Lindbergh) Lake, over Elbow Ridge, and down to a crossing of Glacier Creek, where we started the tedious climb to Lagoon Lake, our camp spot for the night. The air was crisp throughout the day and autumn's touch was noted everywhere along the trail: here and there aspen leaves spiraled gently down to form a yellow carpet beneath the trees; patches of huckleberry bushes were turning to deep crimson; clusters of orange mountain ash berries were eye-compelling when glimpsed against a blue sky. Bird life, too, was evident: crossbills were busy pulling cones apart to reach the seeds, red-breasted nuthatches were calling plaintively to vie with the more emphatic notes of the chickadees. The pine squirrels seemed especially busy, sounding staccato notes of alarm as they scampered away at our approach. It seemed a friendly forest on that bright October day, and best of all we had a feeling of being a part of it—a feeling very precious, to be cherished and long-remembered.

It was almost sundown when we reached Lagoon Lake, and we lost no time in pitching our tent and starting a fire from wood left neatly stacked by the Mountaineer party in August. After a supper of canned beans, cheese, and bread, we sat awhile by the cheerful blaze, grateful to have accomplished one of the longest legs of the trip,—in miles at least. Before turning in we noticed a veil of high cirrus clouds coming from the north to drift across the face of the moon which lacked a few days of being full.

In the early morning hours the noisy flapping of the pup tent disturbed our slumbers. Poking our heads out from the covers to see what it was all about, we were greeted by a blast of frigid air and stinging snowflakes; quite definitely there had come a break in the halcyon autumn weather, at least in the high country. Rearranging our duffle to act as a partial windbreak and making sure our tent was securely anchored, we burrowed beneath the down covers to get a few fitful winks until daylight.

Crawling out of the tent at daybreak we were greeted by a wintry world. The air was filled with blowing snow which eventually came to rest in drifts of considerable depth. The troubled waters of Lagoon Lake slapping the rocky shores could be heard above the roaring of the storm. Definitely this was no day to travel in the high country; even a hot breakfast was out of the question, and to retreat before this onslaught of winter seemed the logical thing to do.

After making up emergency packs containing our down quilts, the axe, and a generous supply of provisions, we cached the remainder of our outfit under the tent, well-weighted down with rocks, and headed for the lower country. In less than an hour we were in a more temperate climate, with the snow turning to rain. However, there was no lessening of the wind, which seemed to be coming out of the north with the fury of a winter blizzard. The mountains were completely obscured by low-hanging clouds; the trees writhed and groaned and frequently there would be a loud crack as some victim fell before the tempest. "No place to be in a windstorm," said Jack. "Nothing to do but hole up until it's over."

And that is just what we did. The Holland Lake cabin provided the shelter we needed, and there we stayed the better part of two days and nights. A captain and his orderly from Fort Missoula shared quarters with us. The deer-hunting season was in full swing and these men were out for game. They were likable chaps, and we enjoyed swapping yarns with them in front of the over-worked little stove.

All the next day it rained steadily, and there wasn't much to do but to stay inside and listen to the steady drum of the drops on the roof. We felt completely isolated—no telephone or other means of contact with the outside world. But we made the best of it, and rather welcomed the opportunity to rest our tired muscles which were not wholly broken in to the hard exercise. Nothing to do but wait, or to wonder perhaps how our outfit left in the high country was faring.

At noon on the third day the storm broke and the sun came out. Jack suggested we start immediately for Lagoon Lake. "We can jungle out tonight somewhere along the trail," he said, "and reach our cache at Lagoon before noon tomorrow." And so equipped with a rusty fry pan, a lard pail, a couple of tin cups, a battered plate or two with some tableware—all of which could be abandoned after use—we started. The captain promised to call Jack's wife as soon as he reached town, and to tell her that our return would be delayed. This he neglected to do, with consequences that will be related later.

We camped that night in a little hollow on the. side of Elbow Lake Ridge in a wickiup made from spruce boughs. Jack located a dead tree which he worked up into firewood, and we kept a fire burning all night, as it turned bitterly cold. Stepping away from the bright circle it made we found the moon had risen to lighten the dark woods. There was not even a trace of wind to murmur among the tree branches. The only sound was the crackle from our campfire calling us back to its cozy warmth. Jack brewed himself a cup of tea with melted snow, and told of happenings when he was building railroads across the Canadian plains. Finally, after piling wood on the fire to last most of the night, we turned in to the luxury of a deep bough bed.

We reached Lagoon Lake before noon as planned, and after digging our gear out of the snow we repacked and headed for the Lake of the Clouds. The way was a hard one; the drifted snow made the footing uncertain, and we proceeded with caution. Minor lakes such as Gem and Prospector were frozen over. Turquoise, a larger tarn, had ice along the shoreline, but was open in the middle. Sunrise Glacier gleamed white with its covering of new snow. We made camp just short of the Lake of the Clouds in a saddle known as Lone Tree Pass, several hundred feet higher than Turquoise Lake.

We took care to make our camp snug as we expected to be there two nights. Again Jack cut down a dead pine, and we were soon at home by our fire. This spot was near timberline, probably at an elevation of seven or eight thousand feet, and the night was bitterly cold. There was just enough breeze across the pass to whisper all night in the stunted pines near our tent. In the morning a kettle of water was frozen solid, and Jack had to thaw it out to get water to mix his famous pancakes.

Hardships were forgotten as we climbed Glacier Peak. The air was crystal clear, and we could see that at lower elevations the day was sunny and warm—a return to Indian summer. Here, high up on the peaks the air was crisp and invigorating. The glare of the sun on the snow was intense. Having no snow glasses, we did the next best thing: rubbed charcoal from the campfire on our faces just under the lower eyelids. This helped, but by evening we both had glorious sunburns.

We crossed one corner of Sunrise Glacier on our climb to the peak. At one place there was a crevasse, too broad and deep to be concealed by the new snow. It had an ugly, repellent look and we avoided it altogether. The warm sun on the cliffs above caused some melting which resulted every now and then in a fusillade of small rocks which went bounding and skipping by us, warning us of incipient danger.

Eventually we reached the crest of the backbone of the Missions, which extended miles to the south. A section of this knife edge forms the headwall of Mission Canyon, a basin on the floor of which are several lakes, not visible from where we were standing. Not far away in a southerly direction was the massive pyramid of Mountaineer Peak. A host of other mountains, including impressive Daughter-of-the-Sun completed this southern panorama. To the north was the huge bulk of MacDonald Peak, a challenge to climb which we accepted the following day.

We took a slightly different route on our descent to camp, passing a snowshrouded tooth which we dubbed "The Ghost." Sitting beside the campfire that evening, thoroughly relaxed and cheered by a warm meal, we laid plans for the next day. We decided to go down to Iceflow Lake, the source of Post Creek, and to follow down that stream to MacDonald Lake and the Flathead Valley.

High cirrus clouds were drifting in from the west and south next morning and we sensed a change in the weather. We lost no time in packing up and picking a precarious way down to the bottom of the abysmal ravine which cradles Iceflow Lake. A more dismal spot would be hard to find—or to imagine. The beetling cliffs of the Glacier Peaks cut off the sunlight except possibly for a few hours each day in mid-summer. The lake was frozen solid that morning.

Looking up at MacDonald Peak, we speculated on how much of a climb it would be, without packs, to reach the top from where we were. Jack thought about an hour and a half would put us on top, and we decided to give it a try. After leaving the lake we climbed westward to a saddle from which we could see the Flathead Valley. Then turning northeast we started up a talus slope—a stiff climb where the footing was insecure because of the new snow, which became deeper the higher we went. Finally we came to the crest of the ridge and found we were some distance from the main summit, which appeared to be a hundred or more feet higher than the point on which we were standing. The wind was now blowing a gale and the sky had become completely overcast. Blowing snow was combing out over a cirque which we thought might hold a small glacier. We could not face the wind standing up, so we crouched as we worked our way along the top of the serrated ridge. At last we were confronted by a gap which could not be crossed with any degree of safety. It is possible that this gap could be crossed easily in good weather, but we did not care to attempt the passage in the face of the wind and snow.

Although we did not gain the summit of MacDonald Peak, we felt repaid for the effort of the climb. Views near and far were ample compensation: close at hand the nearby peaks savage in their wintry aspect; in the distance the white ramparts of the Swan Range; to the west the broad Flathead Valley with its cultivated acres.

Skidding and sliding down the scree slopes and making deep tracks in patches of drifted snow, we reached our packs in a fraction of the time taken for the tedious upward climb. We were soon on our way down the valley below Iceflow Lake. For the first mile or so Post Creek flows in an easterly direction; then it turns north, and finally west to MacDonald Lake.

Two of the most attractive lakes we saw on this trip lie in this high valley. One, which is possibly a half mile across, is situated just north of Panoramic Peak, which has been mentioned before.

It is only a short distance from this lake to another one almost as large. We camped at the lower end from where there was an unobstructed view of Panoramic Peak. The clouds in the west broke away just before sunset and the peak was suffused with a salmon glow which in turn was reflected in the lake. This was one of the richest color effects seen on the entire trip, but it lasted for only a few minutes. We built a rousing campfire from wood picked up along the shore, and cooked a kettle of spotted pup (rice and raisins) which, with cheese and dried prunes to chew on, appeased appetites whetted by the day's exertions. We took stock of our remaining supplies and found that we had food enough for three or four meals. "Close figuring," said Jack. "We may have to live off the country!" "And what a pleasant experience that would be," thought I as I looked across the lake at towering Panoramic Peak, now snowbound until spring.

It started to snow in the night—not a dry snow such as we encountered earlier on the trip, but a wet snow which clung to everything it touched. The falling flakes were so dense that we could not see the lake as we ate breakfast. Packing up our equipment in the cold, gusty wind that was sucking down the valley was quite a chore. Before we had gone far through the brush we were soaking wet from the waist down. Only the exercise of climbing over windfalls and working our way around cliffs kept us from getting stiff and numb.

We passed a number of lakes, one below the other, with cliffy escarpments between. It was necessary to detour in many places and let ourselves down several hundred feet by holding onto the brush. As we dropped in elevation the snow turned to rain, which was little better than the snow and more penetrating. We were soon literally soaked to the skin. It was so cold and we were so wet that it was impossible to stop and rest with comfort, so we kept on the move continually. I shivered most of the time and my knees creaked. Our only comfort was in knowing that each descending step put us closer to our destination and farther from the chill of the snowline.

We left camp about seven-thirty, and it was after noon when we reached the big timber above MacDonald Lake. Here in a clump of western redcedars we found the bones of a horse; the whitened skeleton was one of the most heartening sights of the trip, for it indicated that we were near a trail of some kind—probably the Indian trail which led up the valley and over the divide at the head of the North Fork of Post Creek. We soon found this trail and followed it down to a Reclamation Service camp where the cook cheered us with freshly-fried doughnuts—for Jack there were copious mugs of hot coffee. Below the camp we found a spot where we could build a huge fire. There the rest of the day was spent in drying out.

Next morning was bright and sunny. Trudging along the road towards Ronan we were picked up by the driver of a Model T Ford who was on his way to town. There a friend at the bank loaned us money for haircuts and shaves, and tickets for the afternoon train to Missoula. From the station platform we took a farewell look at the Missions, the tops mostly obscured by angry-looking storm clouds that had settled far down on the great peaks. There was a feeling of wonderful satisfaction in knowing we had made a successful crossing of the great barrier: the challenge of the mountain wilderness had been met and won. That challenge is still there for those who would go on foot into that superb country.

Things had been happening in Missoula, so we found on our return. Because we were long overdue and no word as to our whereabouts had been received, a search party was being organized. The Associated Press had carried a nation-wide story and Jack's relatives living in California had wired for latest reports. But on the whole I think our respective families were less worried than the Forest Service boys. These men were truly concerned, for they well knew the hazards of wilderness travel under wintry conditions.

Towards the end of the summer of 1928, Shoemaker proposed that he and I get pictures in the Missions to supplement those taken by Asahel Curtis—a collection which remained the property of the Northern Pacific Railroad. Ernest Holmes, a teacher at the Idaho Normal College who had been on Montana Mountaineer trips led by Shoemaker, wished to go with us. We were glad to have him as a third member of the party for he was a good man in the hills and always willing to do his share of the work.

The few pictures I brought back from the snowy trip with Jack Clack were made with an Eastman roll film Kodak. As this present trip was primarily for pictures, we decided to carry my favorite piece of equipment, the 6 1/2 x 8 1/2 inch view camera using cut film which I loaded into the holders by means of a changing bag. A sturdy tripod was included in the outfit, which all told weighed about thirty pounds. Shoemaker and Holmes carried a goodly share of this hefty equipment, a courtesy which I certainly appreciated.

At the lodge on Lindbergh Lake we rented a packhorse to take our outfit to Lagoon Lake; once there the animal would be turned loose to find his way home—something he had done several times before. The trip was made without event, and after unpacking the animal we took a swim in the lake, which had been nicely warmed by the summer sun.

Two overnight side trips were made from our base camp: one to the Lake of the Clouds, the other to High Park, where in the evening we watched two grizzlies turning over rocks in a search for ladybugs. Both Shoemaker and Holmes had seen grizzly bears before in several places in the Missions, notably near the top of MacDonald Peak.

This is a land of wind-tortured trees. One could spend days searching out and taking pictures of these sturdy individuals on the timberline frontier. To me they epitomize a gallant struggle for survival. Of the negatives made on the trip, those of the sub-alpine trees were the choicest in my estimation.

Trail-riding parties initiated by the American Forestry Association in the early nineteen thirties made it possible for increasing numbers of people to gain a more intimate acquaintance with wilderness areas in many parts of the country. One of the best of these horseback tours—a trip popular from the first—led into the very heart of the mountain country of the Lewis and Clark and Flathead National Forests now designated the Bob Marshall Wilderness in honor of the man who did so much to give us a greater appreciation of our wildlands and an understanding of the need to protect certain of them from unwise encroachment. Largely through his efforts and generosity the dream of a wilderness society became a reality.

Joe Murphy, well-known outfitter of Ovando, Montana, guided the first Trail Riders' trip into this area in July, 1932. With the exception of the war years, the trip has been repeated annually ever since. However, Joe no longer acts as guide. I joined the party making the ride in 1935 and again in 1946 for the purpose of getting movies and stills which would depict life in camp and on the trail. Total results amounted to many still negatives and a two-reel movie entitled "Trail Riders of the Wilderness" which had a rather wide circulation, particularly in the East. My expenses were taken care of by the Forest Service. Mrs. Swan, although not a member of the Association, went along as a paying guest at the suggestion of Joe, who knew of her years of experience in outdoor living and nature guiding. The Forest Service Remount Depot, located in the Ninemile Valley a short distance from Missoula, furnished me with a saddle horse and a pack mule, which were taken by truck to the jumping off place at the Montour Ranger Station. My mount was a large black horse well broken-in for mountain travel; the mule was a lovable animal of gentle and cooperative disposition. Jack Clack gave freely of his time to make a couple of stout packboxes which could be slung from a Decker saddle. I am reasonably sure that no photographer had ever taken to the hills with better equipment.

A briefing session for the 1935 party was held in the Florence Hotel on the evening before departure. There the members were told what they might expect in the days ahead. The timid ones were reassured; advice on footwear and clothing was given; guidebooks prepared by the Forest Service were handed out.

Next morning before the riders boarded the bus which would take them to the jumping off place at Montour there was a grand flurry of excitement as individuals dashed hither and thither to make last-minute purchases, to mail postcards, or to check surplus baggage which would be cared for at the hotel until their return. The Missoula Mercantile Company had opened early as a special accommodation to the party.

They were fine looking people that finally boarded the bus—men, women, a child or two—all feeling very proud and western in their stiff new Levis, their brogans or riding boots, their multi-colored shirts and jackets. Teachers, secretaries, lawyers, doctors, bankers—these were the people who were to travel together for ten adventure-packed days in the wilderness. And here let me add that seldom during those ten days, in spite of breaking-in fatigue and the petty annoyances of trail travel, did one hear a complaint uttered or a discordant note sounded. It would seem we had been presented with a new set of values when we entered the wilderness.

The Murphy family furnished most of the brawn and muscle for making camp, looking after the seventy or more head of stock, saddling and packing the animals—a gargantuan job, so it seemed to the onlooker. Perhaps success of the expedition may be explained by a remark Joe made as he vaulted into the saddle to take his place at the head of the procession of riders: "This is our vacation. It's a lot more fun than working in the hayfield!"

Tents were provided for those who cared to use them, but sleeping "under the stars" had a fascination for many of the party, and sleeping bags were often unrolled in the open or beneath the trees. Mrs. Swan and I used a small crawl-in tent which protected our equipment from the heavy dews common to mountain meadows.

The horses, who were turned loose some distance from camp to graze during the night, were rounded up shortly after daylight and were driven to a location near camp where they were held in readiness for saddling and packing. It was pleasant to lie for awhile and listen to the thud of the hoofs, the cries of the wranglers, and the notes of a Swiss bell before turning out to yawn and stretch in the morning sunshine. Nobody sleeps late on trips like this. Glances over the bivouac would show numerous heads popping up from sleeping bags with here and there a newly awakened sleeper modestly dressing beneath the covers - a practice much in favor with groups such as this.

Mrs. Swan and I usually ate breakfast with the wranglers so that we could ride out on the trail ahead of the riders to choose positions from which to take pictures of the cavalcade in motion. These early morning hours were choice moments for us: deer or elk might be seen; we were sure to hear the notes of Swainson thrushes, vireos, and western tanagers; foolhens and blue grouse were so common we paid little attention to them. Passing through the deep woods we might hear the trill of a winter wren— a bird easier to hear than to see. High among the branches of the Douglasfirs would be Audubon warblers, tiny birds not hard to identify after one becomes familiar with the characteristic notes.

When we heard from some distance away the chatter of the approaching pilgrims we would take our positions beside the cameras ready to "let them have it" as they passed by. Then later the packstrings would appear, mules and horses stepping along at a lively gait in spite of heavy loads, and often raising a cloud of dust disconcerting to a cameraman. Riding with the packers would be the cook and his assistant, waving to us as they passed. After the dust had settled we would take our places at the rear of the procession until we caught up with the riders at lunchtime. The packstrings did not stop at noon, but proceeded directly to where camp would be made for the night. I generally took advantage of the noon hour to take closeups of the riders eating and lounging. These people seemed thoroughly relaxed and at ease after the first day or two in the saddle, and there was always enough banter and horseplay going on to make things interesting.

To me the most interesting camp was in a meadow at the foot of the Chinese Wall. The Wall is an escarpment of thousand-foot vertical cliffs which for twenty miles forms the east face of the Continental Divide—a tremendous barricade not easily scaled except at one point. From the brink of this remarkable formation one has impressive views both north and south of sheer facets which catch the rays of the morning sun while the valley below is still in shadow. The streams which rise here to flow away to the east are tributaries of Sun River. To the west the land slopes away more gently and is drained by streams which flow into White River, an important tributary of the South Fork of the Flathead River. On one occasion we climbed to the summit. In places there were snowbanks to cross and it seemed a wise precaution to have the party roped together. Some members found this an exciting experience, and an incident to talk about, I am sure, when they returned home.

In season there are fine displays of beargrass (sometimes called squawgrass or elkgrass) in the vicinity of the Wall. The plant is not a grass; it belongs to the Lily family. Because it is mentioned in the Lewis and Clark Journal, it is presumed that these explorers named it.

The upright, conical racemes of small white flowers surmount stems two or three feet high. As the seeds mature, mice and other small rodents climb these stems or chew them off in order to reach this food supply. Some say bears consider the roots edible tidbits when they first come out of hibernation. Mountain goats are known to nibble the stiff, grass-like leaves. But it seems to many lovers of the out-of-doors that the contribution these showy blossoms make to the beauty of a mountain landscape far outweighs these more material values. Thousands of creamy-white plumes gracing a meadow or climbing a slope beneath great peaks are sights to be remembered. In Glacier National Park the beargrass is considered the Park flower.

Before present regulation forbidding private structures in wilderness areas went into effect, Joe had a comfortable cabin on the South Fork near the mouth of the Big Salmon River. Here among other furnishings were two rocking chairs famous in the annals of the locality. "How did you ever get them here?" Joe was questioned by an elderly rider who was enjoying a taste of home comfort after a long day in the saddle. "That was easy," responded Joe. "They came in by packhorse; take out a few bolts and they come apart." Now the cabin is gone—demolished to comply with regulations. But the two rocking chairs are in safekeeping at Joe's ranch in Ovando.

In 1936 the last day's ride took us from camp at Tango Creek to Smokey Creek and thence by the Necklace Lakes to Gordon Pass and down the famous switchbacks to Holland Lake. There we were met by a bus which took us back to Missoula. In 1946 we used a new trail which followed a much easier route, passing Upper Holland Lake and going directly down

Holland Creek to Holland Lake—a fine piece of engineering and badly needed. But those who remember the old trail miss the thrill of descending those agonizing switchbacks with what amounted to an aerial view of Holland Lake spread out below—the familiar range of the Mission Mountains forming the western horizon. Alas, some of us yearn for less progress!

Outstanding among mountain lovers in the early years of the Forest Service was J.C. Whitham, one of those stalwarts who from the first recognized the value of the national forest wildlands for wilderness recreation. Whit, as he was called by his associates, became supervisor of the Gallatin Forest in 1931 and after taking charge lost no time in calling attention to the scenic resources of this remarkable area. Very soon we became acquainted with names that whetted our curiosity: the Spanish Peaks, the Hilgard Area, Specimen Ridge, the Sphinx. With elation I accepted Whit's invitation to join a party which would spend several days—maybe a week—in the Spanish Peaks. There would be several women in the party, and Mrs. Swan was invited to be one of them. "Brother, you must have lived right," commented my laboratory assistant as I took off for Bozeman, the rallying point for the party. There it developed that because of unforeseen circumstances Mrs. Swan was to be the only woman on the trip! The final roundup included Whitham, ranger Arden Gunderson, Mrs. Swan, and myself.

Whitham was an intrepid rider in the hills. "Where a man can tread a horse can go," seemed to be his motto. Up Hellroaring Creek we went from the Squaw Creek Ranger Station to camp at a spot from which was a stunning view of Gallatin Peak. The next day I saw Whit casting glances at this eleven-thousand-foot mountain, and I felt sure he was figuring out a way to get to the summit. Sure enough, that proved to be Iris plan; however, we never made the top. High up on a ridge where we intended to leave the horses and proceed on foot we were caught in a terrific thunderstorm. Donning our slickers we crouched in a spot protected somewhat from the wind by jumbled boulders. Our horses, one of which carried the camera equipment, huddled close together; I remember watching hail bounce off their wet, shiny backs. Lightning struck several times on the ridges above us, and the crack of thunder was ear-splitting. In a matter of minutes the worst of the storm was over and we had the comfort of warm sunshine. At about our level a golden eagle soared gracefully over the valley, riding ascending air currents with almost motionless wings. It became a bright, hospitable mountain world as we made the long descent to camp—the sound of muttering thunder growing less and less in the distance.

Our last night on the trip was spent at Mirror Lake: here we broke camp and packed up for the long trip over Indian Ridge to the Squaw Creek station. As we parted company Whit remarked, "Sometime I must take you to the Hilgard country. I like it as well, if not better than the Spanish Peaks."

Two years later Whitham kept his promise. The place of rendezvous was the Forest Service ranger station at West Yellowstone. Jack Clack joined the party on special invitation, as did Mrs. Swan, who accepted gladly when she learned that Mrs. Whitham was coming, too.

The Hilgard area certainly came up to expectations aroused by Whitham's descriptions. Here we found lakes well-stocked with fish, superb mountains to photograph, well-placed trees to frame choice bits of landscape; meadows with ample horsefeed. Jack's admiration for the country was unbounded, and he spent many hours in helping me choose spots from which to take pictures; the whole trip proved to be an exhilarating experience.

One evening as we were talking by the campfire after a strenuous day, I said in a half-joking manner, "Jack, there is one more world to conquer—the Anaconda-Pintlars." His terse reply set the stage for one of the most interesting trips we ever took together: "Well now, that is a sensible idea; when do we start?" I think at that moment we both would have been delighted to start on the next day if such a move had been possible: as Whit remarked, "This wilderness stuff sure gets in your blood!" As it turned out, several years passed before Jack and I were able to firm up plans for this trip.

The Anaconda-Pintlar Wilderness comprises nearly 158,000 acres of rugged mountain country lying on both sides of the Continental Divide in the Deerlodge, Beaverhead, and Bitterroot Forests in southwestern Montana. Streams on the south side of the Divide flow into the Big Hole River to eventually reach the Missouri; those on the north are part of the watershed which feeds into the Clark Fork of the Columbia. In glacial cirques overshadowed by precipitous and barren peaks are nestled many lakes—lakes which give rise to streams which in their nascent moments meander through meadows rich in mountain vegetation. In some places there are pure stands of alpine larch (Larix lyallii)—stands so clean of underbrush and down timber that one may ride through them seated comfortably in the saddle with never a need to dismount. Elk, mountain goats, and moose are frequently seen, each finding its suitable habitat within the boundaries of this magnificent land.

At the time of our visit few trails had been built in the area. When the supervisor of the Deerlodge Forest learned of our projected trip he called us from Butte to offer full cooperation provided we would do a little exploring to determine possible trail routes. He promised to send rangers Warren Stillings and Eric White to assist. He also told us the Butte Chamber of Commerce was anxious for pictures to use in promotional work, and suggested we include some women in the party to pose in dude clothes. Stillings agreed to bring his wife, Eleanor, provided Mrs. Swan would also join the expedition—a plan to which she agreed without hesitation.

Having obtained official sanction for the trip Jack arranged for saddle horses and a string of blue-ribbon pack mules to be brought over by Forest Service truck from the Remount Depot at Ninemile. Bill Bell was to have complete charge of the mule string, and authority to decide where they could and could not be taken in the rugged country we were to explore. The women, both of whom were experienced camp cooks, would help with the meals. I started taking pictures as we left Storm Lake. Warren and Eleanor, very photogenic in their dude outfits, proved willing subjects for pictures throughout the trip. Kodachrome shots with the Leica made the most of bright neckcloths and western hats against a background of deep blue sky.

On Storm Pass was an ancient alpine larch nicely placed for a picture, and there Jack and I turned our horses over to Bill Bell to be led with the string down to Page Lake where we were to camp. After taking a picture of the larch, we began the descent over a trail that was steep and full of rolling stones. Jack carried the camera and film holders in a packsack; I shouldered the unwieldy tripod, and stuffed the Leica inside my shirt. The pockets of an old vest came in handy for stowing away accessories. It seemed good to be out of the saddle: neither Jack nor I enjoyed riding a horse down a steep mountain grade.

We loafed along, taking pictures as we went. At last we spotted the meadow where we would camp, and noted that Bill had already turned the horses out to graze. When about a half hour later we reached the edge of the grassy park we were met by Bill, who told us with a long face that Fanny, one of the prize mules, was gravely ill—so ill that he didn't expect her to recover. "She's down," said Bill, "and I don't think she"ll ever get up. I've seen them this way before, and they don't often get over it."

This was sad news indeed, for without Fanny we would have to cancel the trip. It was a glum party that gathered around the supper fire. But Fanny did recover. She was lying some distance from our camp, and the two women went over and talked to her after supper. What they said I do not know, but after a few words she struggled to her feet and was soon her old self. After Bill had made a final check he came back with a face all smiles and said, "She'll make it, I do believe. Never saw it happen before. Some of you folks must have done a powerful lot of praying." Perhaps some of us did, who knows?

Taking pictures with a cumbersome view camera such as was commonly used at the time I started my picture-taking career in the Forest Service was a much more laborious process than shooting scenes with the superb and easily handled equipment available today. But work with the old-fashioned view box that had to be mounted on a stout tripod had its compensations. For me there was immense satisfaction in pulling a black cloth over my head and seeing the image on the ground glass viewing screen. Of course, this image would be upside down, but with a little practice the photographer learned to make allowances for this inversion, and to judge the merits of the composition. I often moved the camera several times in composing a picture. Patience paid big dividends in final results I soon learned. Proper exposure depended on the good judgment of the operator. Exposure meters were in the experimental stage when I first started taking pictures. The payoff for meticulous care in the field came in the darkroom when one held up to the light a perfectly exposed negative of pleasing composition!

Jack and I spent many hours together searching for pictures; many little incidents which perhaps at the time seemed inconsequential come back to memory after the passage of years. I remember the morning at Page Lake when in company with Eric White we discovered a smaller tarn pretty much hidden in the timber—a little gem that certainly merited a picture. Our enthusiasm was somewhat dampened, however, when we discovered that a good-sized spruce tree stood in the way at the only point along the shore from which a pleasing view could be taken. Moving the camera didn't help the situation—wherever we stood the tree remained smack in our line of sight. Finally Jack went back to camp and returned with an axe; soon the intruding tree was out of the way. The resulting picture has been used several times in various publications. I chuckle every time I see it, remembering as I do how it was obtained.

Because of the lack of good trails in the Anaconda-Pintlar area at that time the trip proved to be rather hazardous for the mules and horses. In one place Bill led the packstring across a scree slope where the footing was precarious even for the sure-footed mules. We all held our breath as we watched the animals pick their way gingerly across this talus slide, feeling that the slightest earth tremor might send the whole mass of rock tumbling on a downward course. But Bill, expert as he was in dealing with such situations, got his charges safely across and signaled to us to follow. There was no alternative way around this impediment, and so we were obliged to follow the leader, remaining in the saddle and trusting to our horses to step cautiously over the unstable boulders. Safely over at last we could laugh at the little pikas (conies) who squeaked at us from their rocky doorsteps. Some call the little animals "rock rabbits," but their rounded ears resemble in no way the long ears of a rabbit. One often sees on the rocks little piles of vegetation gathered by the pikas and placed to cure in the sun: tiny haystacks which will be safely stored among the rocks for a winter food supply.

Each day as Jack and I were taking pictures, White and Stillings were exploring possible routes for trails, and taking notes to be used in planning a construction program. Days slipped by quickly and time was running out when we reached the headwaters of La Marche Creek. Bill advised that we terminate the trip at this point, and leave the high country to follow a trail that would eventually take us to the highway in the Big Hole Valley. But before turning our backs on this country, we spent a day fishing in a lake where the fish would really bite. Jack and Bill took over the job of frying the catch, while the rest of us sat by the fire and summed up the accomplishments of the trip. We were all satisfied that we had covered thoroughly a section of mountain country which would in the years to come attract many visitors seeking wilderness adventure. Predictions we made that evening by the campfire have proved to be correct: each year increasing numbers of hikers and riders enjoy this superb area designated as the

Anaconda-Pintlar Wilderness. I am glad to have been an early visitor there.

Short trips taken on the spur of the moment with little preliminary preparation often yield pictures of unusual interest. In this category was the trip Jack Clack and I took to the top of Mt. Stuart early in March, 1927, to photograph a surprising display of "snow ghosts." A member of the Lolo Forest told us of this remarkable array of snow sculptures, and we made plans to visit the area on the following day, which was a Sunday. Throughout this particular winter the road to the head of Spring Gulch, where the Mt. Stuart trail starts, had been kept open by woodcutters; this was a good break for us, as it shortened the trip up the mountain considerably.

It was long before daylight when we left the car and put on our snowshoes. As we tightened the straps of the webs we glanced up and noted that the stars were shining in a cloudless sky, promising fine weather for the day ahead; it was with a feeling of elation that we donned our packs and started to climb. Each of us carried a candle lantern improvised from a lard pall—a contraption once known in the Northwest as a "palouser," a term said to have originated in the Palouse wheatlands of Washington, and to have reference to the early rising habits of the ranchers in that locality. In those years a rancher named Lappe had a log home near the head of Spring Gulch; as we passed the dwelling we saw no lights, indicating that members of the family, normally early risers, were still abed.

We didn't talk much in the pre-dawn hours: about the only sound to be heard was the crunch, crunch of the snowshoes. The wavering light from the palousers proved unnecessary in finding the trail, so we left these crude headlights hanging on the brush. Just as the sky was reddening in the east we reached the top of the first ridge to be surmounted. Here the snow was fifteen or twenty feet deep: small trees were completely buried or left with just the tops sticking out.

As the sun came up the whole snowy world was suffused with pink. We had often watched this morning glow on distant peaks; now we felt ourselves an intimate part of the glorious spectacle.

Ahead we could see the top of Mt. Stuart, possibly a half mile away and seven or eight hundred feet above us. From here the final climb to the top took us through a zone peopled with the most outlandish figures one could imagine. Moisture-laden winds from the Pacific Ocean had wrought masterpieces no human sculptor would attempt to duplicate. Wet snow had clung to the branches of alpinefirs and whitebark pines in unbelievable amounts, later to freeze and receive a bombardment of flying ice particles of irresistible force which put the final touches to the strange figures. We saw now that the description given us by the Lolo ranger was not in the least exaggerated.

I titled the first picture I took that morning the "Hall of the Old-fashioned Ladies." At a glance one can see here the pompadour hair-dos and the bustles. At a later time I took Mrs. Swan to the same spot after a heavy snowfall, but the ladies Jack and I saw had changed their attire and lacked the hair-dos and bustles.

*Splendid Was the Trail*

Half a day's exposure to the March sun reflected from the snowfields gave Jack and me deep sunburns which brought color to faces bleached by a winter of desk work. Sun glasses were indispensable, of course. We ate the first of several lunches in the shade of two trees which had been bent together to form an arch. Jack rustled enough pitch to start a fire for his coffee, and we leaned back in comfort on a few boughs he lopped off with the axe. Frankly, I was tired, but the feeling of fatigue was partly overcome by the exhilaration of being surrounded by this world of astounding shapes high above the valley smog.

Photographically the day was highly successful and we left the summit of Stuart at four in the afternoon with a set of pictures which were used extensively in the years which followed: the negatives still show no signs of deterioration.

In subsequent years Mrs. Swan and I climbed Mt. Stuart several times in winter and I got additional pictures to add to the series made with Jack. On one of these trips we ate supper just below the top of the mountain and came down the rest of the way by moonlight. The day had been rather warm, causing the snow to soften near the surface and stick to our snowshoes. Masses of snow would loosen and fall to the ground from the trees with a thud, reminding one to be watchful and stay from under the burdened branches. In late afternoon when the temperature dropped below freezing tiny icicles formed on some of the trees; after sundown a heavy crust made shoeing hazardous. Several times we skidded on the hard surface and had nasty falls; finally we sat down and slid—a maneuver which wore through the seats of our heavy wool pants. Eventually we reached home that evening just before midnight—and called it a day!

Photographing game animals is a fascinating pastime which calls for a lot of patience and some knowledge of the habits of wild creatures. I have never enjoyed hunting with a gun, but I have spent many hours stalking game with a camera. If I were asked to name my favorite wildlife study I would probably mention a moose picture which I took one Labor Day weekend at a little Idaho lake near Elk Summit in the Selway-Bitterroot Wilderness. Ed Mackay, then in charge of the Powell Ranger District, told me that several of his men had seen a moose feeding in this lake on several occasions—a report which seemed to spell picture possibilities. Ed gave me encouragement in the idea: "Early morning is the most likely time to see the animal—get out before breakfast," said he. Accordingly I drove to Elk Summit in the evening so as to be on hand early the next morning.

Leaving the car I took my position in a clump of lodgepole pines near the lake shore, prepared to wait until something happened. It was a cool morning and I was grateful to be warmed by the sun when it came above the ridge across the lake. The only sound I remember was the sharp rat-tat-tat of a pileated woodpecker sounding off on a snag somewhere in the distance.

An hour passed and nothing happened. I about made up my mind to leave in another five minutes; had I done so I would have missed one chance in a thousand to get the picture I wanted. Just as I was unscrewing the long lens from my Graflex I heard a cracking of branches across the lake and a large bull charged out of the timber, plunged into the shallow water of the lake, and started to feed on vegetation growing on the bottom. The next half hour proved to be one of the most exciting I ever spent with a camera. My subject didn't seem to mind my being in the vicinity, and I got closer and closer without disturbing him. It was not until I had used all available film that he changed his mind and started toward me in what seemed a rather threatening manner; it was then I gave ground and made for the car as quickly as possible. Back in the darkroom that evening I found that this Labor Day had been a very profitable holiday for me: I have never been able to get better pictures of this largest member of the deer family.

# The CCC: Adventure for Youth

The Civilian Conservation Corps came into being as a nation wide project in 1933. From its inception it met with an immediate response from thousands of youths in all parts of the country who wished to avail themselves of the opportunities it offered for useful employment and training in various skills. The work which these boys accomplished under the guidance of competent instructors and foremen was truly impressive.

From the very first the Forest Service was deeply involved in this program. Soon the need was felt for a comprehensive pictorial record of work being done by the Corps on the forests in all parts of the country, and the work of getting the pictures was turned over to several photographers working for the Forest Service. Each man was given a certain territory to cover: mine included all of Region One, the Black Hills, Wyoming, Colorado, Utah, Oregon, and western Washington. Both motion pictures and stills were asked for. To cover this vast area and to get satisfactory pictures in six weeks—the time allotted for the job—proved quite a problem. I was on the road early and late, and several times in order to check results I turned a hotel room into a laboratory and worked far into the night developing films.

My scheduled tour ended abruptly when the great Tillamook fire broke out in Oregon and I was sent there to get pictures. Many of these shots fitted nicely into the CCC series as they showed boys on the fireline. Sometimes the scenes were very spectacular, with plenty of smoke and flame in the background.

Most of the foremen on the line were old experienced hands in fighting fire. It was gratifying to note how well they handled their crews of green boys, and how solicitous they were for their safety—there were no accidents on the fireline during the time I was taking pictures. This assignment was a rather grueling experience but I was grateful for it; once home, it was something to look back on with satisfaction. Many of the movie scenes I took were used extensively.

In the ensuing years I visited many camps and took many pictures, both stills and movies. One camp which I remember was located on Priest Lake in northern Idaho. It was under the jurisdiction of the Kaniksu Forest. Here the boys were engaged in such work as tree planting, road and trail maintenance, blister rust eradication, campground improvement, and numerous other jobs.

On one weekend Mrs. Swan and I were enjoying a vacation at our favorite campground on Luby Bay in sight and sound of this camp. It was Sunday evening, and after a day mixing with the holiday crowd of picnickers we sat by our fire enjoying the view across the lake to the Selkirk Range, now rosy in the last rays from the setting sun. Two boys from the camp, wandering rather aimlessly along the lake shore, noted our fire and stood looking our way. Seeing they were hesitant to intrude, we invited them to sit in the smoke which afforded some protection from the hordes of mosquitoes that had put in an appearance as evening approached. They were likable chaps who soon overcame their shyness and told us that this was their first trip away from the pavements of New York City. Asked how they liked our country, one of them replied, "It sure is big, and there ain't much to listen to," and looking across the lake to the deepening blush on the mountains added after a long pause, "Pretty, ain't it!" Before taps our newly made friends went back to their barracks. They were Negro boys from Harlem!

109                                    *Splendid Was the Trail*

Many of the enrollees I talked with in other camps were boys with country backgrounds—boys used to caring for farm animals and driving teams of horses. Others were handy with tools and rendered excellent service in building roads and trails. Under the direction of skilled foremen, cadres of boys took part in building ranger stations and other structures on the forests. They proved to be excellent workers in forest nurseries where stoop labor was needed for weeding seedbeds and transplanting young trees.

Early in the CCC program the Forest Service felt a responsibility in providing a little frosting for the cake in the way of occasional motion picture showings in the camps—programs which would not only entertain but instruct as well. A man to develop this activity showed up at just the right time: Elmer Bloom, an old newspaper man, tactful, a good organizer, and with a flair for photography. Elmer soon had a fleet of showboats on the road, manned by boys from the camps who had been given special training in driving a truck, operating and caring for motion picture equipment, and keeping necessary records. They were an elite force, these boys working under Bloom's direction, and needless to say popular with the boys in the camps they visited.

My farewell pictures of the Civilian Conservation Corps were taken on a frosty October morning at Lolo Pass on the Montana-Idaho line when the Packer's Meadow Camp disbanded and the boys climbed aboard trucks which would take them to Missoula where they would entrain for their homes. Ranger Ed MacKay was there to give them an hilarious send-off by tossing out sacks of smokechaser rations from the back of a pickup. The ensuing scramble was something to see—a bunch of exuberant youngsters, tanned by a summer of Idaho sunshine, scuffling for the last time en masse. For some of these boys work in the CCC had opened up vistas of new opportunity. As for me, I was grateful to have had the chance to record on film some phases of the program.

# Recruits for the Home Front

During the years of World War II, finding labor for the many necessary jobs on the home front became a major problem. The Forest Service and other Federal agencies found their manpower badly depleted as a result of men leaving for military service. Farmers found it difficult to get men to care for and harvest their crops. Anyone depending on hired help often found himself scraping the bottom of the barrel.

During this period the Forest Service was very grateful for the help

*Splendid Was the Trail*

afforded by the women who accepted jobs as fire lookouts. Up to this time it was unusual to find a lookout manned by a woman, but during this period of stress and strain it became apparent that a woman could handle this work fully as well, if not better, than a man. Now lady lookouts are met with on many forests.

"Get pictures of the gals on the peaks," said my chief after calling me into his office. "There's a story there to be told, and pictures will be needed in the telling. Take Mrs. Swan with you, if she will go, and spend enough time with these women to get a complete story—little incidents as well as pictures to dress up the tale." Suffice to say that Mrs. Swan jumped at the chance to go along as chaperone, and we were soon on our way in the "Silver Streak," an International station wagon with a top silvered by a coat of aluminum paint. We carried with us, in addition to the camera equipment, bedrolls and supplies so that we could camp at the lookouts when necessary. If you haven't gazed up at the night sky from a bed spread on a mountain top, you have missed a pleasant experience!

Many of the women we visited were schoolteachers, and in talking with them we found that, far from being lonesome, they enjoyed being by themselves in a romantic setting of mountain country. Some had made friends with mantled ground squirrels and bird visitors. Packers usually visited the lookouts at fairly frequent intervals, taking up supplies, mail, and sometimes water in ten-gallon cans.

Water on a mountain top is a luxury. Often, a lookout had to get her own supply by daily trips to a spring or water hole quite a distance down from the summit. Conversations with the ranger or dispatcher over the telephone in the evening seemed to be part of the daily routine. In the evening each lookout called the dispatcher at a stated time to make an official report, and afterwards it seemed that a little gossip about the day's happenings would be perfectly permissible. As one girl put it, "I don't get lonesome here in the daytime, but in the evening it's sure nice to know what is going on below!" Rangers—in fact, all personnel—seemed to be doing everything possible to make each lookout's summer a pleasant and safe experience. Not once on our tour did Mrs. Swan and I hear anything but praise from lady lookouts for the treatment they were receiving from forest officers.

While eating breakfast at home in Missoula one morning I got a call on the telephone saying, "Hurry down to the Northern Pacific Depot. A trainload of Mexican Nationals has just come in. The men are being sent up the Bitterroot to work for the beet-growers. You should get pictures of them."

By the time I reached the station the men had left the coaches in which they had traveled for a couple of days and were lined up for registration prior to taking trucks and buses to various points in the Bitterroot Valley. They were a goodnatured bunch of men, happy at having an opportunity to earn what seemed to them fabulous wages. The morning air was chilly and it was not long before they were given blankets—where obtained I don't remember—as protection against the unaccustomed cold. I think all of us who gathered at the station that morning were glad to give a cordial welcome to these likable braceros. Later, while waiting for the final beet harvest, many of them worked for the Forest Service piling brush and doing other work for which they were qualified.

Harvesting sugar beets is a big chore. Lacking sufficient experienced labor for the job, some growers employed teenagers to help with the work. High school students in Ronan, St. Ignatius, and other places in the Flathead Valley were excused from their classes during short periods for this purpose. In company with a reporter from a local Missoula paper, I made a trip one afternoon to get pictures to illustrate a newspaper story. The crews we visited were made up of husky ranch boys and girls who did a fine job topping the beets prior to loading them for shipment to the sugar factory in Missoula.

At one time a need arose for posters dealing with the proper preparation and storing of food as part of the war effort, and I was asked for suitable pictures. The offices were canvassed for photogenic subjects, and those selected assembled one afternoon in the Montana Power Company's kitchen to be photographed. It was a most pleasant assignment.

A search for rubber substitutes was made during the war. Guayule, a shrub belonging to the Composite Family, seemed to be a possible source, and large plantations were started in California. At the end of the war when natural rubber was again available in increasing quantities, the project was dropped. Another plant, koksaghyz, commonly known as Russian dandelion, was grown in an experimental way in the Target Range area near Missoula. Weeding and picking of the blossoms was done mostly by women. The experiment was carried to a successful conclusion, proving that rubber could be obtained from this source, but at a cost prohibitive except in dire emergencies. At our request, two Russian technicians were sent by the Soviets to inspect the plantation and give us advice on cultivating the plant. These men spoke no English, and we felt fortunate to have a native-born Russian on our staff to act as interpreter: Serge Skoblin. an engineer who had lived in this country for many years.

For a considerable period, several hundred interned Italian seamen were cared for at Fort Missoula. Many were well-educated and talented men who had held positions of responsibility on luxury liners. Members of ships' orchestras gave concerts and entertainments still remembered by those in the community who were fortunate enough to obtain tickets. Many townspeople who had crossed their fingers when the men first came changed their attitude and expressed regrets at seeing them leave at the end of the war. I for one gained a better understanding of Italian culture and resourcefulness by mixing with these people. From coal passers and cabin boys right up the line to first officers, I found them friendly and dependable. Drawing on this considerable pool of manpower, the Forest Service was able to do much on the forests in the way of brush-piling, road and trail maintenance, building repair, and the like. They were seamen, not woodsmen, but they did a good job for us. I often wished I might know how they fared when they returned to their homeland.

The Kootenai Forest found it expedient to employ a crew of brush-pilers composed mostly of women to do a cleanup job near the Ant Flat Ranger Station, a work center not far from Fortine. I took pictures of this project near the very end of the war. The work was being supervised by several oldtimers who had worked many years for the Forest Service. It was gratifying to see how efficiently and tactfully they supervised their crews—composed almost entirely of female workers, many of them teenagers.

These foremen and a few other summer employees ate their meals at the Forest Service mess, which was presided over by a motherly person who had left a lovely home in Kalispell to do her bit for the war by serving as a cook during the summer. Her teen-age daughter was bull-cook and flunky. This refined and lovely person did her best not only to provide delicious meals but to improve the deportment of those who sat at her table. One soon noticed that little niceties of table manners were being observed, banter was at a minimum, and uncouth remarks entirely lacking. A climax was reached when an old woodsman friend of mine, well-versed in lumber-camp customs, was admonished not to drink from his saucer. After the meal he said to me as we crossed the yard, "She sure knows how to cook, but doggone it, she tried to teach us old guys etiquette!"

# Project: Photographic Points

We know that in the course of events slight matters are often caught up by the imagination and enlarged to a size out of all proportion to their importance. Such was the case in the incident of the writer's famous hat. In the early twenties this hat appeared in a picture of a group of white pine trunks, where it had been hung to indicate the size of the boles. Some twenty years later the same hat—a little battered perhaps by the vicissitudes of many years in the woods—was shown in the same spot in a follow-

*Splendid Was the Trail*

up picture. The coincidence was mentioned casually in hearing of the editor of the Forest Service news sheet who, being short of items at the time, gave heed to the incident and used it as a space filler.

The response from the field was immediate and hearty, even to offers of financial assistance in obtaining a new headpiece. And so this old stiff-brimmed Stetson, which cost me $5.00 in the first place, probably did more to stimulate interest in Project Photographic Points than many a circular letter sent to the field by the regional office!

What is a photographic point? Briefly stated, it is an established point, carefully chosen, from which pictures can be taken at intervals to show changes that have occurred during the time lapse between shots. These pictures are often of great value in showing the growth of plantations, the establishment of reproduction on old burns, the changing aspects of timber sale areas, the rate of decay of stumps or fallen timber. It is a fascinating project, and the finding of spots from which original pictures had been taken (with no thought at the time of a follow-up) gave me more thrills than any other branch of forest photography during my years with the Service—except, possibly, the taking of wilderness views.

In 1909 the much publicized Lick Creek timber sale was in progress on the Bitterroot National Forest. This operation, which was intended as a sort of pilot project for future sales, was in the charge of W.W. White, one of the pioneers in the early years of timber management. I suspect that during the progress of this sale he spent most of his waking hours among the trees that he loved. Certain it is that he lived in a cabin on the area and became well acquainted with every aspect and feature of the topography in the vicinity. The Big Blackfoot Milling Company, a subsidiary of Anaconda, purchased the timber and did the logging according to Forest Service specifications. For years the sale area was considered a sort of show window of conservative cutting practices and logging methods.

At this time a Forest Service photographer named Lubkin was sent out from Washington to take pictures of the sale area. In those days the standard camera for such work was a 6½- x 8½-inch view box using glass plates. This man got an excellent series of pictures which have been used extensively throughout the years. But no record of the points from which the views were made was kept at the time, as far as anyone knows.

During November of 1925 White and I took a set of the original pictures and went to the sale area to see if we could locate the points from which they were taken. The quest was extremely fascinating. White had a good memory and was able to spot, in a general way, the locations we were after. Peculiar stumps and logs were a great help. Just when we might seem baffled in the search for a particular spot something would show up to give us a key. The clue might be the bark pattern on a ponderosa pine, or perhaps a forked trunk.

The camera we were using duplicated the one used for the original pictures, and when a spot was once found it was a simple matter to adjust the outfit so that the image on the ground glass would coincide with the print we were holding. It was an exciting game, and we felt it was more fun than work. In a few days we had located all but one of the points—sixteen in all. Later the points were marked by iron posts which were tied in to a Land Office corner by compass and pacing. The traverse was platted on a large scale map for future use.

In the winter of 1923 I had the opportunity of taking pictures of logging by the Beardmore outfit which was working at the time on the Kaniksu National Forest north of Priest River, Idaho. In those years winter transportation in that country was by horse and sled. Car travel, off the main highways at least, was unthought of. And so when I met Howard Drake, then in charge of timber sales on the Kaniksu, he told me Hank Ogston, government scaler on the job, would meet us with the mules after lunch at the Priest River Hotel. That he did. The mules were hitched to a small sled, and after wrapping ourselves in various robes and sugans we started on the 18-mile drive to the notorious old Tunnel Camp of the Beardmore operation. The name "Tunnel" came from the fact that the main building was long and tunnel-like, with few windows. Few woodsworkers of the present day would live and work under the conditions then found in many of the camps.

Logging on this sale was by trail chutes; strings of logs were hauled along the chutes by horses to a terminus a few chains above camp. Here they were loaded by a "jammer" to sleds which carried them over an iced road to decks on the bank of the West Fork of Priest River to await the spring drive. The top-loader at the cross haul was the king of the outfit—a Paul Bunyan whose bellow could be heard far back in the woods. Chain saws were not dreamed of then; trees were felled with crosscut saws pulled by husky Swedes who often worked stripped to the waist.

For two days Drake and I tramped over the sale area on snowshoes, taking pictures between snow squalls. It was gloomy with never a spot of sunshine. Only as we were ready to leave did the sun come out to give me a chance for a shot of Hank and the mules headed for town. The pictures I got on the trip reflect the dark weather.

I returned to the area the following July, at which time brush was being piled. It was on this visit that I hung my hat on a small branch protruding from the trunk of a white pine seed tree and took the picture which was to attract so much attention in later years. It was the Fourth of July, to be exact, and the brush pilers had gone to town to celebrate. I recall how peaceful it seemed working alone in the deserted woods, fragrant in the hot July sun. I didn't pay much attention to the locations from which I took a series of pictures that day. Why should I? It never occurred to me that I might return for follow-up views at a later time.

Just previous to World War II, I was given a definite assignment to locate as many points as possible from which pictures had been taken on this old Beardmore sale, mark them with stakes or monuments of a permanent nature tied into established corners or enduring landmarks, and retake the original views. The aspect of that bit of country had changed greatly since

I had taken the first series of pictures twenty years before. Reproduction had been good on the old sale, and in places formed dense stands. Trying to find identifying objects in this jungle of new growth was a frustrating experience. For a couple of days I looked for clues without success.

John Murray, long a scaler on the Kaniksu, solved the puzzle after following some 40 miles of old chutes, mostly on snowshoes. One Sunday as he walked up the draw in which lay the rotting remains of a main chute, he noticed a feeder chute coming in from the side. The mouth of the draw in which it lay was almost obscured by young trees and would hardly attract attention except to the most careful observer. Proceeding up this draw (we later named it Hidden Draw), John came into a broad, basin-like swale rimmed with gently sloping ridges. Here were the missing points— almost all of them. Here was the tree on which I had hung my hat twenty years before!

During the war, travel was greatly restricted and forest officers used whatever vehicles they were able to commandeer. Assigned to me was an International station wagon of ancient vintage—a hard-riding dust trap which, nevertheless, gave good service during the years I used it. I gave the top a coat of aluminum paint to turn the sun's rays; hence the car was usually referred to among Forest Service men as "Swan's Silver Streak." Into this wagon went all sorts of equipment: camera boxes, tools for digging post holes, iron posts, a kapok bed. Alas! A few years ago I found its remains in a wrecking yard on the outskirts of Missoula.

I finally marked the points Murray had found and took the necessary pictures. This happened in May 1944. Not far from the old sale area was famous old Camp 164, originally built for CCC boys but occupied during the war by interned Italian seamen and a few German war prisoners. The cooks at this camp had served as chefs on luxury liners which were then languishing at piers in New York harbor. The men were friendly and eager to talk of their families whom they had not seen since the war started. Many who were skilled craftsmen donated their free time to help build a Roman Catholic church in the town of Priest River. Forest personnel loved to stay at Camp 164 where we all sooner or later acquired a taste for Italian cooking. Here it was that I made my headquarters while retaking pictures on the old Beardmore sale.

Roy Phillips, then supervisor of the Kaniksu Forest, was very much interested in this picture-point project, and soon appeared in camp to give me help. At the Falls Ranger Station was a supply of white-painted wooden posts, and these were made available to me for marking points. These posts, 4 inches square and 6 or 7 feet long, were heavy, and how to get them back in the woods away from roads became a problem. After talking with the camp superintendent, Phillips had four Italians detailed to help. Appointed to supervise the cadre was a captain who spoke good English, as well as four or five other languages, and told us wild tales of running the blockade in the White Sea. These recitations were timed to coincide with post-hole digging, which operation could be directed conveniently from a comfortable sitting position at the base of a tree. The details of the captain's stories are half-forgotten, but I still have a vivid recollection of the trip from camp to the area; the captain beside me on the front seat while three Italians rode the load of posts in the back. The Silver Streak wrenched and groaned over rotten corduroy and through the mud holes as we approached the site of the old Tunnel Camp, but finally made it. Then each of the three men in the back shouldered a post and did some more groaning as we hiked up the old chutes and through the mosquito-infested brush.

Time will determine the worth of this project. In another fifty years or so, some of these pictures may prove valuable beyond all expectations. Who knows?

# Still a Trail to Follow

    This evening as I write I look out of my study window to the grassy slopes of a mountain which rises fourteen hundred feet above our valley. It is a mountain of moods and a calendar of the seasons which holds our interest throughout the year: in spring we eagerly watch for the white blossoms of service berry bushes far up its sides; later children scramble up the slopes to gather showy wildflowers which bloom before the hot, dry summer days; in autumn before snow comes to the valley we scan the mountaintop for the first crown of frosty white. This evening the whole westward-facing slope is bronzed with rich light from the setting sun. Soon the homes at the base of the mountain will be in shadow, but the eastward-bound evening plane will be in full sunlight as it flies high overhead to disappear quickly beyond the mountain wall.

At the close of the last ice age a lake, known to geologists as Lake Missoula, filled this and other tributary valleys to a depth of many hundreds of feet; at the present time one may see on this and other mountains in the vicinity horizontal lines which mark the ancient beaches of this prehistoric lake. This body of water was formed by an ice jam which dammed up the Clark Fork River at a point many miles downstream from Missoula.

Except for a few solitary ponderosa pines, some of great age, and a scattering of bushes with here and there a few younger trees, vegetation on the west side of our mountain is confined mostly to grass. In spite of precarious footing, horses find pasture there, and elk and deer drift down occasionally from the higher mountains in winter to graze in full view of the townsfolk. Once in early September we spotted a little black bear feeding on chokecherries. But settlement takes its toll of wild creatures; we miss the chorus of the coyotes commonly heard at night in past years.

A few steps west of our home is Rattlesnake Creek, which has its source in a series of glacial tarns in the mountains north of town. No one seems to know how the stream got its name—certainly rattlesnakes are seldom, if ever, found along the creek or on the surrounding hills; children carry on their quest for wildflowers with never a thought of reptiles. The Indians had a more attractive name for our creek—they called it "The Stream of the Salmon Trout." During times of high water the little river is a noisy stream; in mid-summer when surplus water is used for irrigating crops its voice diminishes to the merest murmur. In all seasons the flow supplies our city's domestic needs.

A bit of wild land lying on both sides of the creek was donated for a city park by a pioneer resident, Thomas Greenough, who stipulated that it be kept perpetually in its primitive state. Here are great cottonwoods and ponderosa pines, with an understory of shrubs typical of the locality— wildrose, chokecherry, ninebark, syringa, snowberry, hawthorn, and others. The area shelters many species of birds: song sparrows, vireos, redstarts, and others. In winter the leafless trees come alive with chicadees, nuthatches, and juncos, and flocks of Bohemian waxwings swoop down to gather what berries they find. Here in the snowy woods we listen for the sweet song of the Townsend solitaire, and watch water ouzels diving into the icy creek in search of food. Here in our backyard we feel we have a wilderness!

Henry David Thoreau says in "Walden," "As long as I can enjoy the friendship of the seasons, I trust that nothing can make life a burden to me." Mrs. Swan, my companion on many days spent in the out-of-doors, and I both feel ourselves truly blessed by such a friendship; our trail through the Forest Service years and right down to the present time has indeed been a splendid one—could we follow it again we would ask for few changes!

# Index

*Splendid Was the Trail*